Florencio J. Perez. All Rights Reserved.

of this book may be reproduced, stored in a retrieval system, or transmitted by any without the written permission of the author.

published by AuthorHouse 02/07/05

ISBN: 1-4208-1381-1 (sc)
ISBN: 1-4208-1382-X (dj)

brary of Congress Control Number: 2004099148

rinted in the United States of America
Bloomington, Indiana

This book is printed on acid-free paper.

Evolution of Consciousness
A Spiritual Journey Guide to Universal Law

by

Florencio J. Perez

1663 LIBERTY DRIVE, SUITE 200
BLOOMINGTON, INDIANA 47403
(800) 839-8640
WWW.AUTHORHOUSE.COM

Acknowledgement

I want to acknowledge those who helped me complete this book. Special thanks go to my cousin Hernan Contreras who provided proof reading, editing, and much appreciated comments to help with the book. I also offer my thanks to my young artist/illustrator Ashley Wedekind for the front cover painting for the book.

I thank my sister Amalia Perez for relentlessly encouraging me to continue to write the book even when I was dismayed by the length of time it was taking to write it. My wife Susan should be commended for her patience while I spent hours at the computer, and for listening to me talk about the book over her first cup of coffee in the morning. She provided invaluable help by proof reading the manuscript many times and provided corrections to so many of my typos and misspelling.

I would also like to dedicate this book to my two sons and one daughter whom I am very proud of and to their families whom I love very much. They are my son David Perez, and his wife Gwen, and daughter Molly, my son Michael Perez, and my daughter Patricia Perez Watchitzki, her husband Hans, and daughter Nina.

Table of Contents

Acknowledgement v
Forward xvii
Acronyms xix

Chapter 1 An Overview 1
   1.1 The Universal System 2
      1.1.1 Major Components of the Universe 2
      1.1.2 Universal Domains 3
      1.1.3 Building Blocks of the Universe 4
   1.2 Universal Laws 4
      1.2.1 Law of Causes and Effect 5
      1.2.2 Law of Relativity 6
      1.2.3 Law of Evolution 6
   1.3 Universal Spectrums 6
   1.4 Stages of Consciousness 7
   1.5 Components of the Human Being 8
   1.6 Mind 8
      1.6.1 Life 9
   1.7 A Brief Definition of God 10
      1.7.1 Attributes of God 13

Chapter 2 The Human Composition 17
   2.1 The Human Physical Body 18
      2.1.1 The Large Human Brain 19
      2.1.2 Computer Scientists Speculation 20
      2.1.3 Genetics and Cloning 20
      2.1.4 Vital Organ Subsystems 21
   2.2 Astral Body 22
   2.3 Vital Energy 24
   2.4 The Human Instinctive Operating Mind (HIOM) 26
      2.4.1 Computer Analogy 27
      2.4.2 The Trainable Mind 28
      2.4.3 HIOM and Brain Interaction 29
   2.5 Personality 30

    2.5.1 Childhood Environment 31
    2.5.2 Lower Emotions from the HIOM 32
    2.5.3 Past Life Experiences 32
    2.5.4 Soul-Consciousness Contributions to the
    Personality 34
    2.5.5 Molding by the Present Personal Environment 34
    2.5.6 The Changeable Personality 35
    2.5.7 Heredity 35
2.6 The Intellectual Mind 36
    2.6.1 Chain of Command 37
2.7 Intuitive Mind 39
    2.7.1 Brainwashing 40
    2.7.2 Access to Higher Realms 41
    2.7.3 Genius 41
    2.7.4 Personal Experience 42
    2.7.5 Meditation Technique 43
2.8 The Consciousness 43
    2.8.1 Desire and Will 45
    2.8.2 The Consciousness is Subconscious 46
    2.8.3 Analogy of the Automobile 46
2.9 The Soul 47
    2.9.1 Function of the Soul 47
2.10 The Spirit 48
    2.10.1 Analogy of the Sun Shining on Dewdrops 48
    2.10.2 Analogy of a drop of Water from the Ocean 49

Chapter 3   Waves 51
3.1 Introduction 51
    3.1.1 The Most Fundamental Wave 52
    3.1.2 Many Types of Waves 53
3.2 Sound Waves 54
3.3 Electromagnetic Waves 55
    3.3.1 Segments of the Electromagnetic Spectrum 57
    3.3.2 Broadband 58
    3.3.3 Coexistence of Frequencies 59
3.4 Matter and Energy Waves 60
    3.4.1 Atomic Structure 60

    3.4.2 Quantum Mechanics..61
    3.4.3 Excerpt from Schrödinger's Lecture 63
    3.4.4 Subatomic Particles 64
    3.4.5 Antimatter..65
    3.4.6 Dark Matter and Dark Energy 66
    3.4.7 Super-string Theory..66
  3.5 The Mental Universe..67
    3.5.1 The Wavelet Hypothesis..68
    3.5.2 Metaphor of a Wavelet 69
    3.5.3 Properties of Wavelets...69
    3.5.4 Building Material of Mind 70
    3.5.5 Building Material of Consciousness 72
    3.5.6 Building Material of Spirit and Soul.....................72

Chapter 4   The Law of Cause and Effect...73
  4.1 Law of Cause and Effect..75
    4.1.1 Chain of Command for Action 75
    4.1.2 The Human Mind and Cause and Effect 76
    4.1.3 Imaginary Story of a Little Girl and a Hot Grill ...78
    4.1.4 Cause and Effect of Inanimate Things 79
    4.1.5 The Laws of Motion...79
    4.1.6 Cause and Effect from Beginning of Time to End of Time...80
  4.2 The Game of Life..80
    4.2.1 Actions are the Result of Thought.........................80
    4.2.2 Work for the Sake of Work...................................81
    4.2.3 Work, Success, Wealth and Happiness.................84
  4.3 Sub-Laws of the Law of Cause and Effect 85
    4.3.1 "Law of Attractions" stated as "Like Attracts Like"..86
    4.3.2 "Law of Like Conditions" stated as "Like Produces Like"..88
    4.3.3 "Law of Increases" also stated "As you sow, so shall you reap"...89
  4.4 Veil of Ignorance..90
    4.4.1 A Narrative on the "Veils of Ignorance"...............91
  4.5 The Cycle of Human Life 93

  4.5.1 Reincarnation 94
  4.5.2 Leading Events towards Rebirth 97
  4.5.3 Reincarnation Answers Nagging Questions 98
 4.6 Spiritual Cause and Effect 100
  4.6.1 Relationships 103

Chapter 5 Relativity 107
 5.1 Introduction to Classical Relativity 107
  5.1.1 Einstein's Special Theories of Relativity 109
  5.1.2 Einstein's General Theory of Relativity 109
  5.1.3 Facts of Relativity 110
  5.1.4 Motion in the Astral and Spiritual Domains 113
  5.1.5 Einstein's Mental Exercise (The Imaginary
  Train) 113
 5.2 Relativity in Ordinary Things 114
  5.2.1 Money is Relative 115
  5.2.2 Relativity of Words 116
  5.2.3 Relative Humidity 117
  5.2.4 Imaginary Political Spectrum 117
 5.3 Truth 119
  5.3.1 Wealth and Happiness 123
 5.4 Reality 123
  5.4.1 The Affects of Modern Physics on the Perception of
  Reality 124
  5.4.2 Relative Reality 125
 5.5 Good, Bad, Right, and Wrong? 125
  5.5.1 A Good Universe 129
 5.6 Ethics and Rules of Conduct 131
  5.6.1 Spiritual or Religious Revelations 132
  5.6.2 Intuition 134
  5.6.3 Human Reason and Human Made Laws 135
  5.6.4 Right Action 136
 5.7 Relativeness of the Consciousness 137
  5.7.1 Scales of Nature 138
  5.7.2 Relativeness of the individual Human
  Components 139
 5.8 Relativity of Domains 141

Chapter 6 Evolution 143
  6.1 Cosmic Evolution 144
    6.1.1 Origin of the Universe 144
    6.1.2 Contraction of the Universe (the Black Hole) 146
    6.1.3 Evolution of the Earth 147
    6.1.4 Chronological Evolution of Life on Earth 148
    6.1.5 Chronological List of Milestones 149
  6.2 Darwin's Theory of Evolution of Life on Earth 150
    6.2.1 Principles of Darwin's Theory 151
    6.2.2 Organization of Living Organisms 152
    6.2.3 Five Living Kingdoms 153
  6.3 Man Made Evolution 154
    6.3.1 Evolution of Science 154
    6.3.2 Evolution of Tools and Technology 155
    6.3.3 Parallels between Nature and Television 156
  6.4 Nature's Guiding Intelligence (NGI) 157
  6.5 The Metaphysical Account of Evolution of the Universe 160
    6.5.1 The Metaphysical Big Bang 160
    6.5.2 The Metaphysical "Black Hole" 162
    6.5.3 Illustration of Systems Engineering Design 163
    6.5.4 Illustration of a Playwright 164
  6.6 Evolution of the Instinctive Operating Mind 165
    6.6.1 Relationship of Universal Mind and Universal Law 166
    6.6.2 Evolution of MIOM 167
    6.6.3 Evolution of TIOM 167
    6.6.4 Evolution of GIOM 168
    6.6.5 Evolution of PIOM in the Plant Kingdom 168
    6.6.6 Evolution of AIOM in the Animal Kingdom 170
    6.6.7 Evolution of Human Instinctive Operating Mind (HIOM) 172
  6.7 Evolution of Consciousness 173
    6.7.1 Parallel Evolution of Physical Body and Consciousness 174
    6.7.2 Evolution of Plant Group Consciousness 175
    6.7.3 Animal Group Consciousness 175
    6.7.4 Analogy of Dewdrops to Clarify Group

  Consciousness 176
6.8 Evolution of Human Consciousness 177
  6.8.1 Launching the First Humans 178
  6.8.2 Explaining the New Human Parts and Functions 179
  6.8.3 Speech 182
  6.8.4 The First Human 182
  6.8.5 Super-men evolving into Super-Gods 183

Chapter 7 The Physical and Astral Domains Of the Universe ... 185
7.1 The Physical Domain 187
  7.1.1 Planes of the Inanimate Physical Domain 187
  7.1.2 Planes of the Plant Kingdom 189
  7.1.3 Planes of the Animal Kingdom 191
  7.1.4 Planes of the Humans 192
  7.1.5 Planes of the Soul Consciousness Stage 194
  7.1.6 Correlation of the Planes to Physical Evolution .. 194
7.2 Astral Domain 194
  7.2.1 Planes of the Astral Domain 196
  7.2.2 The Inanimate Planes of the Astral Domain 197
  7.2.3 The Plant Planes of the Astral Domain 197
  7.2.4 The Animal Planes of the Astral Domain 197
  7.2.5 The Human Planes of the Astral Domain 198
7.3 Vital Energy 198
  7.3.1 Vital Energy in Humans 199
  7.3.2 Human Auras 200
7.4 Thought and Thought Dynamics in the Astral Domain ... 201
  7.4.1 Effects of Good or Evil Thoughts 203
  7.4.2 Problem Solving 204
7.5 Human Astral Body 205
  7.5.1 Sleep Separation 206
  7.5.2 Astral Projection or Out of Body Experience 206
  7.5.3 The Near Death Experience 208
  7.5.4 The Process of Death 210
  7.5.5 The Process of Death in the Physical Domain 211
  7.5.6 The Process of Death in the Astral Domain 212

Chapter 8   The Spiritual Domains 217
   8.1 Spiritual Domain 1 218
      8.1.1 Evidence of Spiritual Domains 219
   8.2 The Slumber Plane and the Awakening 221
      8.2.1 Awakening in SD1 222
      8.2.2 Examples of Differences upon Awakening in SD1 223
   8.3 Planes of Spiritual Domain 1 228
      8.3.1 SD1 has no Planes for Plants, Animals or Inanimate Matter 228
      8.3.2 Planes for the Human Consciousness Stage 228
      8.3.3 Causation Planes 230
      8.3.4 Barrier above each Plane 230
      8.3.5 Analogies to Explain Barriers between Planes 230
      8.3.6 Living in the Higher Planes of SD1 231
      8.3.7 Living in the Lower Planes of SD1 231
      8.3.8 The Planes of Hell or Purgatory 232
   8.4 Characteristics of Life in SD1 235
      8.4.1 Diversity 235
      8.4.2 Old Age 236
      8.4.3 Dimensions 237
      8.4.4 Family 237
      8.4.5 Clear Thought 238
      8.4.6 Desire, Will Power, and Strength of Thoughts 239
      8.4.7 Thought communication 240
      8.4.8 Money, Knowledge, and Education 240
      8.4.9 Companionship in the Spiritual Domains 240
   8.5 Principle of Reincarnation 241
      8.5.1 Evidence of Reincarnation 241
      8.5.2 Reincarnation 243
      8.5.3 Final Preparation and Influences for Rebirth 244
      8.5.4 Dying in SD1 245
      8.5.5 Final Rebirth and the New Life on Earth 246
   8.6 Spiritual Domain 2 247
      8.6.1 Planes of the Bottom Half of SD2 247
      8.6.2 The Commencement Sub-stage 248
      8.6.3 Planes of Multi-Sensory Sub-stage 248

    8.6.4 Planes of Physical Zenith Sub-stage 248
    8.6.5 Planes of Etheric Sub-stage 249
    8.6.6 Planes of the Top Half of Spiritual Domain 2 249
  8.7 Characteristics of Life in the Top Half of SD2 250
    8.7.1 Love in the Top Half of Spiritual Domain 2 250
    8.7.2 The Creative Instinct 251
    8.7.3 Mental Imaging 251
    8.7.4 Dimensions 252
  8.8 Spiritual Domain 3 252

Chapter 9 Consciousness 255
  9.1 Plant and Animal Stages of Consciousness 255
    9.1.1 Plant Consciousness Stage 255
    9.1.2 Animal Consciousness Stage 257
  9.2 Mechanics of Transition from Animal to Human 259
  9.3 Human Consciousness Stage 262
    9.3.1 Sub-Stages of the Human Consciousness Stage 263
    9.3.2 Sub-stage 1 – New Human 264
    9.3.3 Sub-stage 2 – Early Human 266
    9.3.4 Sub-stage 3 – Primitive Human 267
    9.3.5 Sub-stage 4 –Intermediate Human 267
    9.3.6 Sub-Stage 5 – Civilized Human 268
    9.3.7 Sub-stage 6 – Refined Human 270
    9.3.8 Estimated Distribution of Sub-stages in our World Today 271
  9.4 Soul Consciousness Stage 272
    9.4.1 Sub-Stages of the Soul Consciousness Stage 273
    9.4.2 Sub-Stage 1 – Commencement Sub-Stage 273
    9.4.3 Sub-Stage 2 – Multi-Sensory Sub-stage 275
    9.4.4 Sub-Stage 3 – Physical Zenith Sub-Stage 277
    9.4.5 Sub-Stage 4 – Etheric Sub-Stage 279
  9.5 Other Worlds 280
    9.5.1 The Odds of Contacting Other Worlds 285
    9.5.2 Life 285
  9.6 Cosmic Consciousness Stage 286
    9.6.1 The lower Emotions are converted to Love 288
  9.7 Spiritual Consciousness Stage 289

    9.7.1 Characteristics of People of the Spiritual
    Consciousness Stage 289
    9.7.2 Angels 290
    9.7.3 Why Don't They Show Themselves to Us? 292
    9.7.4 Region of Absoluteness 293
    8.7.5 Final Union with God 293
9.8 Analogy of Imaginary Conveyer Belts 294
    9.8.1 Plant Consciousness Stage 295
    9.8.2 Animal Consciousness Stage 297
    9.8.3 Conveyer Belt for Human Consciousness Stage 298
    9.8.4 Conveyer Belt for Soul Consciousness Stage 298
    9.8.5 Conveyer Belt for people of the Cosmic
    Consciousness Stage 298
    9.8.6 Conveyer Belt for people of the Spiritual
    Consciousness Stage 299

Chapter 10   Techniques for Spiritual Growth 301
    10.1 Four Methods or Techniques for Growth 302
        10.1.1 Ignorance 302
    10.2 Right Living 303
        10.2.1 Hard Work 304
        10.2.2 Clean Living 305
        10.2.3 Devotion to Family and Community 306
    10.3 Education and Training 308
        10.3.1 Education 308
        10.3.2 Metaphysics 310
        10.3.3 Training 311
        10.3.4 Benefit of Hi-Tech 311
    10.4 Love and Devotion to God 312
        10.4.1 Evolution of Religion 313
        10.4.2 The Advanced Conception of God 317
        10.4.3 Prayer 319
        10.4.4 Love 320
        10.4.5 An Example of Praying for Love 321
        10.4.6 Law of Attractions 322
        10.4.7 Praying for help from your Guardian Angel and
        Spiritual Guides 323

- 10.5 Mind Control..325
    - 10.5.1 The "Silva Mind Control Method"......................326
    - 10.5.2 Meditation and Levels of the Brain and Mind ..327
    - 10.5.3 Reaching the "Alpha Level" 329
    - 10.5.4 Converting Negative Emotions to Positive Emotions..330
    - 10.5.5 Emotional Awareness 331
    - 10.5.6 Becoming Aware of the Emotional Average 332
    - 10.5.7 Back to Meditation 334
    - 10.5.8 Phobias 336
    - 10.5.9 Bad Habits 337
    - 10.5.10 Benefits..338
- 10.6 Awareness of Consciousness..339
    - 10.6.1 Tools of the Consciousness 341
    - 10.6.2 Computer Analogy 342
- 10.7 Mastery of the Mind...343
    - 10.7.1 Programming the Intellectual Mind 344
    - 10.7.2 Application of Affirmations.................................345
    - 10.7.3 Illustration for Developing an Awareness of Self 348
    - 10.7.4 Acceleration towards Soul Consciousness Stage 350

Bibliography 351

Forward

This book combines the knowledge of the latest modern day science with the knowledge of metaphysics to explain a philosophy of being. Modern day science provides a very good understanding of the workings of the physical universe which is a creation of God. Therefore, knowing the physical world is like knowing a very important segment of God's creation. Science confines itself to understanding the physical universe from a strictly objective standpoint. Metaphysics, on the other hand, is a philosophy that studies both the physical as well as the nonphysical universe. Metaphysics studies the subjective or unseen part of the universe which is a vast segment of God's creations. The Western World tends to shy away from metaphysics because of past dogmas that suppress these philosophical studies. This book emphasizes metaphysics as a means for understanding the hidden subtleties of God as well as understanding our own Spiritual Being. In actuality, God does not hide anything; everything is wide open for our inspection; we just have to grow in understanding to be able to see it all.

The scientific information for this book is obtained from modern day scientists or writers on the subject of physics such as Richard Morris, Roger S. Jones, and Stephen Hawking. Much of the information is also available in many well known encyclopedias such and World Book, Encarta, Britannica, and others. The scientific content in this book is intended to be understood by the layperson; it is at the level of most popular encyclopedias of today.

The origin of the metaphysical knowledge ultimately comes from Consciousnesses greater than us. Two basic sources of metaphysical knowledge for this book are, 1) Yogi Masters of India, and 2) Modern Mediums or Spiritualist that have access to highly advanced Consciousnesses in the Spirit World. The bulk of the Yoga information comes to us from Yogi Ramacharaka who wrote a series of books in the early part of the Twentieth Century. His books are an English translation of the teachings of the Masters of India who themselves got that knowledge from higher Spiritual Sources. The books by Yogi Ramacharaka were initially intended as a series of lectures that were delivered in America in the early 1900s. The content of these lectures were then printed in the form of a series of books which are listed in the bibliography.

Another important source of metaphysical knowledge comes to us from the numerous modern day Mediums or Spiritualists. Spiritualists are able to make contact with Consciousnesses that are more advanced than us and as a result they provide us with invaluable information. Two highly regarded Mediums are Edger Cayce and Ruth Montgomery. They get their information while in a meditative trance and later relate the information to us through transcriptions of those sessions. The objective is making contact with Consciousnesses that are truly more advanced than us. There is no advantage to communicating with Consciousnesses that are at the same level as us since we already know what they know.

Acronyms

AI	Artificial Intelligence
AIOM	Animal Instinctive Operating Mind
AM	Amplitude Modulation
CD	Compact Disk
CPA	Certified Public Accountant
CPS	Cycles Per Second
CPU	Central Processing Unit
DNA	Deoxyribonucleic Acid
DVD	Digital Video Disk
EEG	Electroencephalogram
ESP	Extra Sensory Perception
FM	Frequency Modulation
GIOM	Galactic Instinctive Operating Mind
HIOM	Human Instinctive Operating Mind
IR	Infrared
MBA	Masters in Business Administration
MCC	Mission Control Center
MIOM	Mineral Instinctive Operating Mind
NASA	National Aeronautics and Space Administration
NGI	Nature's Guiding Intelligence
OS	Operating System
PDA	Personal Digital Assistant
PIOM	Plant Instinctive Operating Mind
RAM	Random Access Memory
REM	Rapid Eye Movement

ROM	Read Only Memory
SD1	Spiritual Domain 1
SD2	Spiritual Domain 2
SD3	Spiritual Domain 3
SETI	Search for Extra Terrestrial Intelligence
TIOM	Terrestrial Instinctive Operating Mind
UHF	Ultra High Frequency
VHF	Very High Frequency
VLF	Very Low Frequency

Chapter 1

An Overview

The universe and the laws that govern it were created for Evolution of Consciousness. Consciousness starts with the most primitive forms of life and gradually, over millions of years, advances in sophistication to the level of the modern human and beyond. In the process of development, Consciousness goes through a series of stages using a multitude of tools that have been setup to make this development possible. The universe itself is the primary tool that was created by God the Absolute to support the evolutionary process. The universe is an intricate system with many components that work in unison in an orderly, perfect, and harmonious fashion. It is a hierarchical system that works flawlessly like a perfectly designed machine. It is the goal of this book to explain the basic workings of this complex system and its contribution to Evolution of Consciousness. The method used to describe the process is to first explain the known traits of our universe using the latest scientific knowledge and then infusing the known traits with philosophy, theology, and metaphysical reasoning to complete the descriptions. The explanations start with the objective and ends with the subjective. It is a technique used to tie the known scientific knowledge with the metaphysical philosophy to explain a multifaceted subject in a rudimentary way.

1.1 The Universal System

The universe is a *"system"* composed of many parts or components. System parts are described and discussed independently, but they are all interrelated and operate as one unit. It takes all the parts to make up the universe; all parts are essential and absolutely necessary for the universe to work as a unit. If any one part were to be removed, the universe would not work. American Heritage Dictionary defines a system as **"*A group of interacting, interrelated, or interdependent elements forming a complex whole*"**. Because all the components of the universe are so interrelated, it is difficult to write about any one component without continually making reference to other components that are to be covered in later chapters. This brief overview of the contents of the book is provided to give the reader a prospective that should help connect the pieces as the discussion progresses.

Current thought is that the universe is basically a Physical Universe. It is thought to consist of matter and energy in intergalactic space with life confined to our planet Earth. American Heritage Dictionary defines the universe as *"All matter and energy, including Earth, the galaxies and all therein, and the contents of intergalactic space, regarded as a whole"*. The universe is much more than that. This book holds to the proposition that there is one universe that is composed of five domains, and that God, who is Absolute, created this relative universe. The universe was created in God's Infinite Mind and manifested out of God's Mental Substance. The beginning of the relative universe was the beginning of relative time and space, which occurred at the time of the "Big Bang". The latest estimate from the scientific community is that the "Big Bang" occurred about 13.7 billion years ago.

1.1.1 Major Components of the Universe

Refer to Diagram 1 for a general representation of the universe. Remember that from our prospective the universe is infinite, and a one-page diagram of the universe in not much, but it will serve to provide a general idea. The block at the top of the diagram represents God who is Absolute and is the creator of the relative universe. In essence God forms the universe in His Infinite Mind and then manifests it into existence. This implies that the universe is made of God's Mind Substance, which means that our universe is a mental universe. And, it must be noted that God's Mind Substance is as real as real can be.

The Universal System is extremely complex and elaborate; it fits together perfectly and operates faultlessly. All components of the universe are important, and there are no extra or superfluous components; every component is absolutely necessary to make the System operate without flaws. The explanations presented in this book regarding the universal components and their operations are basic and elementary, but sufficient to give the reader an essential understanding. The universe is hierarchical in nature and the lists of universal components provided are arranged in a way that reflects this order. The lists normally show the least significant component at the bottom and the most significant component at the top. The Universal Laws are all considered to be of equal values, so there is no order of significance in this case.

1.1.2 Universal Domains

As mentioned previously our universe is a single universe with five domains. In other words, we do not have multiple and parallel universes; instead we have one universe with multiple domains. The Physical Domain is the only domain that humans at our stage of development are familiar with, and it is the lowest domain in the hierarchy of the universe. The Physical Domain is the equivalent to what we commonly know as the Physical Universe. Spiritual Domain 3 is the highest and most advanced domain, and it is the domain that we know least of all. Domains are further subdivided into planes, which are also hierarchical, and will all be explained in detail in Chapters 7 and 8. The domains in descending order are as follows.

5. Spiritual Domain 3 (SD3)
4. Spiritual Domain 2 (SD2)
3. Spiritual Domain 1 (SD1)
2. Astral Domain
1. Physical Domain

1.1.3 Building Blocks of the Universe

The most fundamental of all building blocks in the universe is mind, energy, and matter. Matter is made of compacted energy per the equation $E = Mc^2$, and energy is made of compacted mind, but up until now, there is no equation for this principle.

- Mind – There are billions of levels of mind; mind exists in every Domain and every Plane within each Domain.
- Energy – Energy also exists in every domain and every plane within each domain.
- Matter – Physical matter exists only in the Physical Domain and astral matter exists in the Astral Domain. There is no matter in the Spiritual Domains.

1.2 Universal Laws

Universal Laws permeate, influence and dominate every part of the universe. They influence every moment of every day of our lives and are present and active in every nook and cranny of every domain of the universe. Universal Laws are absolute and do not change; they are the same everywhere all the time, they are unbreakable and they are equal and impartial. They permeate all matter, energy and mind. They influence the physical world as well as the Spiritual.

There are many other important sub-laws in the universe, but only a few are treated in detail in this book. Generally the laws of physics, chemistry, biology, geology, astronomy, etc. are not considered Universal Laws because they only apply to the Physical Domain. The "Law of Attractions" is an important law that permeates all, and is a sub-law of the Law of Cause and Effect. It is treated

Diagram 1
God and the Universe

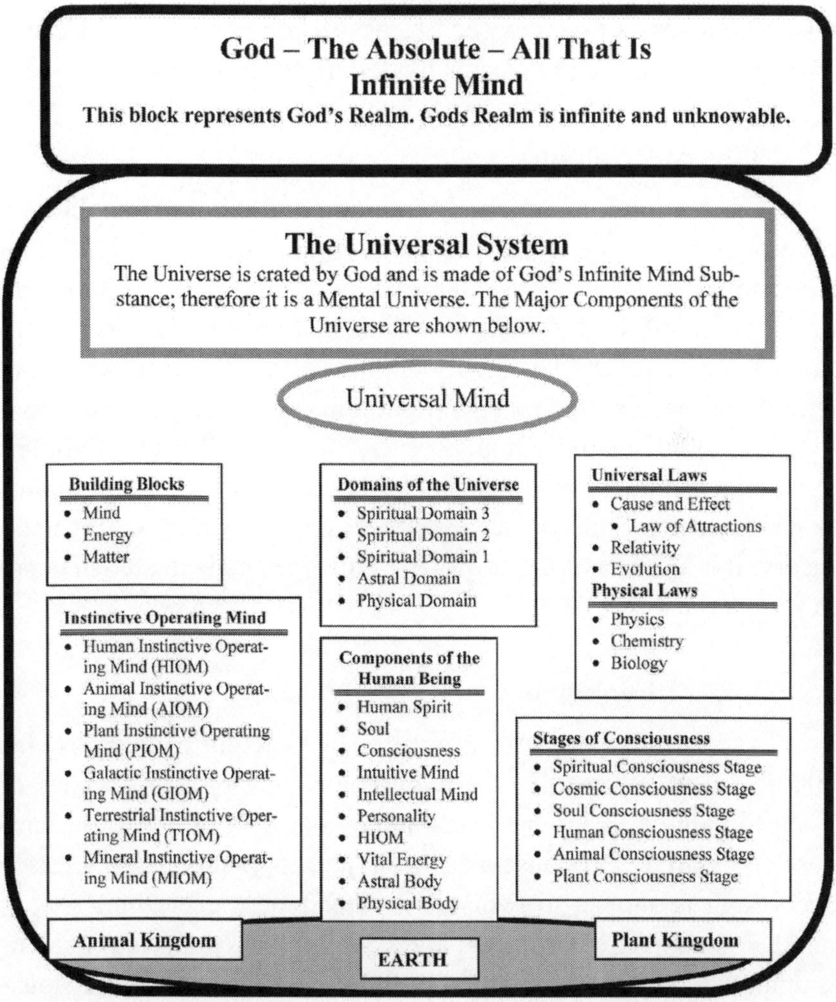

God – The Absolute – All That Is
Infinite Mind
This block represents God's Realm. Gods Realm is infinite and unknowable.

The Universal System
The Universe is crated by God and is made of God's Infinite Mind Substance; therefore it is a Mental Universe. The Major Components of the Universe are shown below.

Universal Mind

Building Blocks
- Mind
- Energy
- Matter

Domains of the Universe
- Spiritual Domain 3
- Spiritual Domain 2
- Spiritual Domain 1
- Astral Domain
- Physical Domain

Universal Laws
- Cause and Effect
  - Law of Attractions
- Relativity
- Evolution

Physical Laws
- Physics
- Chemistry
- Biology

Instinctive Operating Mind
- Human Instinctive Operating Mind (HIOM)
- Animal Instinctive Operating Mind (AIOM)
- Plant Instinctive Operating Mind (PIOM)
- Galactic Instinctive Operating Mind (GIOM)
- Terrestrial Instinctive Operating Mind (TIOM)
- Mineral Instinctive Operating Mind (MIOM)

Components of the Human Being
- Human Spirit
- Soul
- Consciousness
- Intuitive Mind
- Intellectual Mind
- Personality
- HIOM
- Vital Energy
- Astral Body
- Physical Body

Stages of Consciousness
- Spiritual Consciousness Stage
- Cosmic Consciousness Stage
- Soul Consciousness Stage
- Human Consciousness Stage
- Animal Consciousness Stage
- Plant Consciousness Stage

Animal Kingdom　　**Plant Kingdom**

EARTH

in more detail in Chapter 6 along with the Law of Cause and Effect. Reincarnation is essential to the operation of Evolution of Consciousness at the level of the Human Consciousness Stage and it is treated as a principle rather than law.

1.2.1 Law of Causes and Effect

The Law of Cause and Effect is the primary law that guides and propels us through the evolutionary process. It guides us forward whether we know it or not, and whether we like it or not. It is constantly active in every domain of the universe and it does not play favorites. It is equal to all races, all cultures, all economic levels and classes, all geographic places, all religions and all beliefs. The Law of Cause and Effect is treated in detail in Chapter 4.

1.2.2 Law of Relativity

Our universe is a relative universe and relativity applies everywhere in varying degrees according to the Scale of Evolution of Consciousness. Albert Einstein introduced us to relativity at the turn of the twentieth century, but his relativity applies to a narrow section of our existence. His theory applies to the Physical Domain at speeds approaching the speed of light. This book proposes that varying degrees of relativity applies everywhere at all time. It applies to us every moment of every day of our lives yet we are oblivious to its existence. From the standpoint of this book relativity is not a theory; it is a hard and fast law of the universe and is treated in detail in Chapter 5.

1.2.3 Law of Evolution

Charles Darwin first introduced us to evolution in the 1860s. His theory is confined to the evolution of the physical bodies of living organisms. This book maintains that everything in the universe is relative and consequently subject to change and therefore everything is subject to evolve. In other words everything that is relative is subject to evolution and equally everything that evolves is relative. Of paramount importance is the Evolution of Consciousness, especially the evolution of Human Consciousness. This Law of Evolution is treated in detail in Chapter 6.

1.3 Universal Spectrums

Everything in the universe is made of waves; this includes matter, energy, mind and Consciousness. The most fundamental wave in the universe is the *sine wave* and it is treated in excellent detail in High School Trigonometry. A basic understanding of wave theory is essential to understanding how the universe works since everything in the universe is made of waves. In our Physical Domain we are familiar with the electromagnetic spectrum, but the universe contains many other wave spectrums of much higher frequencies. We will be referring to wave frequencies or wavelengths throughout this book. The basics of wave theory are explained in Chapter 3. A few of the major frequency spectrums of the universe are listed below in descending hierichical order.

5. Spiritual Spectrum
4. Mind Spectrum
3. Vital Energy Spectrum
2. Astral Spectrum
1. Electromagnetic Spectrum

1.4 Stages of Consciousness

Consciousness, along with Spirit and Soul, forms the Real Self of each individual. Consciousness is a real thing; it is made of Universal Mind Substance, and is akin to mind, but it is not mind in the same sense; it is the animating substance in the Human Being. This book also treats Sub-human Consciousness, which consist of Plant Consciousness, and Animal Consciousness. It has taken millions of years to reach the stage of Human Consciousness, and there will be millions of years before humans reach the highest stage of Consciousness, which is Spiritual Consciousness Stage. The main thing that can be assured is that Consciousness will continue to advance, grow, and evolve from lower to higher levels, and then from higher to even higher levels and stages, until it reaches the final stage. The final and highest possible levels of advancement is when Consciousness merge with God. The stages and level of Consciousness will be referred to throughout the book but will be discussed in detail in Chapter 9. The following list shows the preferred names for the different stages of Consciousness; listed in the usual hierarchical (descending) order.

6. Spiritual Consciousness Stage

5. Cosmic Consciousness Stage
4. Soul Consciousness Stage
3. Human Consciousness Stage
2. Animal Consciousness Stage
1. Plant Consciousness Stage

1.5 Components of the Human Being

There are ten components that make up the Human Being. The least significant and lowest component of the human is the physical body, and of course, it is the component that we know best. Medical science has made great strides in understanding the physical body in recent years but little is known about all the components above the physical body. This book provides a basic explanation of all ten components in Chapter 2. The list below shows the ten components in the usual descending hierarchical order.

10. Spirit
9. Soul
8. Consciousness
7. Intuitive Mind
6. Intellectual Mind
5. Personality
4. Human Instinctive Operating Mind (HIOM)
3. Vital Energy
2. Astral Body
1. Physical Body

1.6 Mind

Mind is prevalent and pervasive throughout the universe. The highest of all minds is God's Infinite Mind, which is unknowable and resides outside of the universe. God's Infinite Mind created the universe as well as Universe Mind, which is the mastermind. All subordinate minds of the universe operate under the supervision of Universal Mind. Instinctive mind operates the animate as well as the inanimate parts of the Physical and Astral Domains. The instinctive minds in descending order are as follows.

- Human Instinctive Operating Mind (HIOM)
- Animal Instinctive Operating Mind (AIOM)
- Plant Instinctive Operating Mind (PIOM)
- Galactic Instinctive Operating Mind (GIOM)
- Terrestrial Instinctive Operating Mind (TIOM)
- Mineral Instinctive Operating Mind (MIOM)

Human Mind is referred to as Human Personal Mind, and it is composed of the Human Instinctive Operating Mind (HIOM), Personality Mind, Intellectual Mind, and Intuitive Mind. Chapter 2 gives a brief description of the functions of each of these minds.

1.6.1 Life

Life is normally thought of in biological terms, which are physical in nature. American Heritage Dictionary defines life as: *"a. the property or quality that distinguishes living organisms from dead organisms and inanimate matter, manifested in functions such as metabolism, growth, reproduction, and response to stimuli or adaptation to the environment originating from within the organism. b. The characteristic state or condition of a living organism."* In essence, biological science suggests that life is made of biochemical materials such as hydrocarbons, amino acids, proteins, nucleic acids (which forms DNA and RNA), and energy. This book proposes that all life is sacred and is made of physical matter, astral matter, energy, mind, and Spiritual components. A brief explanation of each as it applies to this book follows:

- **Spiritual** – The Spiritual consists of the Consciousness, Soul, and Spirit. These three components are present in all life.
- **Mind** – All life contains a mental component. It can vary from the extremely primitive instinctive mind to the very advanced human mind.

- **Energy** – Energy exists at every level of the composition of life. It exists in the Physical, the Mental, and the Spiritual Domains.
- **Physical**– The physical consists of all the ingredients normally called for in biology. The Physical in this book normally means both the Physical Body and the Astral Body because they are so closely related and interconnect. Both are normally inseparable throughout the lifetime of the person. They separate permanently only at the so-called physical death of the person. There will be more details on the subject in later chapters.

1.7 A Brief Definition of God

This book is not a dissertation on theology of God, but because God is referenced so much throughout, it is proper to provide a definition of God as it applies to this book. It should be noted the term God is not important; every language and many religious faiths have their own word or term for God. What is important is the definition. Many religions faiths or theologies have definitions of God that are vary different in detail but are more or less in agreement of the underlying concept of God. Still there are also many myths and superstitions.

This book contends that God is infinite, and we are trying to define an infinite God with a finite Human mind and a finite vocabulary. The result has to be that our definition of God can never be complete. The best we can do is approach the true definition to the best of our ability. Because our Consciousness is changing, our idea of God is also changing in accordance with our stage of development. God is Absolute and consequently does not change; it is only our conception of God that changes as we develop. As we get wiser our conception of God becomes truer. In other words, our definition of God changes for the better, as we slowly but surely get smarter.

There are many simplistic myths or misunderstandings about God. An attempt is made here to try to clear up some of those misconceptions. For instance, God is not a man or a woman. God does not wear a white robe, He does not have a white beard, and He does not live up in the sky on some cloud called Heaven. The concepts of God that depict Him like a man and having human emotions, passions, habits, and characteristics are called anthropomorphic. This concept of God belongs to the infant stages of the human evolution. The human concept of God has evolved along with human evolution, and rightfully so. In the primitive sub-stages of Human Consciousness, such as the Cave Man or the Stone Age man, the human concept of God was very low. Humans from the very beginning had a built-in instinct to believe in a Divine Being that was greater than them, and even though they knew not what to worship, they did the best they could in accordance with their development. Primitive savages would worship stone idles or some man made idle, or the weather (rain), or the moon, or the sun, etc. They have their gods of war, of peace, of love, of agriculture, of trade, and what not. They worship and try to pacify these various gods, not realizing that underneath it all they are obeying the religious instinct that will in time lead them to worship the One, the Absolute. They worship those things that, in their primitive view, affected their lives and which they do not understand.

The early Greeks and Romans worshiped many Gods. They had Gods for every human frailty, emotion, fear, or joy. As the Human Consciousness evolved, the concept of God changed and became more sophisticated. At present, the prevailing idea of God in our society varies widely. One of the most prevalent beliefs is that God is like a policeman who is watching over us to see that we do not misbehave, and if we misbehave He is ready to punish us. Most of these so-called sins are those that the religious organization defined as sins and they are usually designed to benefit their own agenda. These people tend to always pray for mercy from their God who is ready to pounce on them for wrongdoing. The general belief is that God is a being outside of nature that is greater than the worshiper. This God is merciful if you behave according to the laws of the church and vindictive if you don't.

From the more scientific point of view, it seems that God created the universe with all its necessary laws and then somehow left it to its own devices to continue to operate. The general thought is that Nature now runs itself by the laws of physics that were somehow set in operation a long time ago, and it runs in an automatic mode much like a mechanical timepiece. Unless we understand something about the nature of God, we cannot understand anything about the nature of the universe or of Life.

The term "God" is commonly used in the English language to refer to the highest concept of Deity or Divinity. The term God is also used throughout this book, but it must be noted that other terms may also be used to expand on the higher conceptions of God. Other terms that are commonly used in this book are, God the Absolute, The All, All That Is, The Infinite Living Spirit, Infinite Mind, Spirit, The Creator, etc. The name doesn't matter; names are something we come up with, and every language and every society has its own naming conventions. What is important is the definition that we come up with, and that definition should be of the highest conception that we can possibly imagine. People should also realize that we are evolving beings because we are relative beings, and because we are relative, our concept of God will change as we change due to our evolution.

We are only expected to do the best we can in regard to our definition of God. We can only approach the best definition possible depending on the level of our development. As we evolve our definition will also evolve. God is infinite and for us to try to define infinity with a finite mind and a finite language is impossible. A preliminary definition of God is that God is Good, God is Infinite, and God is Absolute. The bulk of God is outside of our universe, but He created the universe out of his own Mental Substance. Therefore, our universe is of God, that is, it is made of God's Mental Substance. The universe is of God but it is not God. A simple analogy used to illustrate this principle is a drop of ocean water; the drop is of the ocean and like the ocean (as far as the dissolved mineral content is concerned) but the drop is not the ocean.

1.7.1 Attributes of God

Many thinkers of the past including theologians, metaphysicians, philosophers and scientists have been compelled to think that there is something that caused the universe and life into existence. Theologians tended to think that this something is God; the metaphysicians called it mind, reality, truth and similar names; philosophers called it life, or substance; scientists call it matter and energy. In all cases they were looking for the Ultimate Thing, the Creator, the Absolute. Thinkers are obligated to think of, and to search for that underling something beneath it all. The Dutch philosopher Spinoza once wrote "To define God is to deny Him." meaning that God is infinite and it is impossible for our finite mind and language to fully describe Him. God is unknowable to us; He is Absolute and absoluteness cannot be described in terms of the relative. Defining God is speculation on our part, but it is assumed that we should at least honestly try to the best of our ability, knowing that as we advance, our definition will change. The following axioms are a compilation of many ideas by thinkers of the past on the subject. These are axioms that the intellect can accept after some deliberation and contemplation.

1 - God the Absolute IS. This suggests that God the Absolute is the ultimate being. There can be no other being comparable to God the Absolute. The "Absolute IS" means that He is all that can possibly be, and that is the total of all there is. It is complete and there cannot be two whole and complete things. There is only one God who is complete in every way.

2 - Whatever really is must be God the Absolute. There cannot be two or more Absolute Beings. There can only be One God who is Absolute. All other things that appear real to us must be relative. God the Absolute is the only being that really is and anything else that is must be of the Absolute.

3 - God the Absolute is made up of all that really is, all that there has ever been, and all that can ever really be. God the Absolute is all; it is whole and complete; there cannot be anything else but the all; there is nothing that can exist outside of the all. The

Absolute never had a beginning and will not have an ending so it includes all that is, all that has been, and all that will ever be.

4 - God the Absolute is Omnipresent. God the Absolute is present everywhere at the same time. The All must be everywhere, and there cannot be a place with nothing in it. What we call space, time, matter, energy, and minds are but relative manifestations of the Absolute. All matter, energy and minds are made of God's Infinite Mind Substance therefore God is in everything and He is everywhere simultaneously.

5 - God the Absolute is Omnipotent. God posseses all energy and all power that there is. He is all-powerful. All energy in the Physical Domain, the Astral Domain and in the Spiritual Domains is a manifestation of God. It is a relative manifestation of the Absolute. All energy is of the Absolute and there can be no other power. All energy from the sun, from fossil fuels, from nuclear reaction or whatever, is all relative expression of the All.

6 —God the Absolute is Omniscient. God the Absolute is all knowing, all wise, knows everything, and is infinite knowledge. God knows and understands all, and He cannot make mistakes or change His mind for a better idea; if there was something God did not understand or if He made mistakes then He would not be absolute or omniscient and that is impossible. God knows all the past, present, and future and knows it all to the finest detail. God is Infinite Mind, which is absolute, as well as Universal Mind, which is relative. Our human Personal Minds are also manifestations of the mind of the Absolute who knows all there is to know about us. In Mathew 6.8 Jesus was talking about prayer when He said, "for the Father knows what you need before you ask it". This quote refers to the fact that God the Absolute knows all including our own personal mind and thoughts. Nothing can escape God.

7 - God the Absolute is Infinite. Infinite means that God is unlimited or boundless in time and space. He is without limit in power, capacity, intensity, or excellence. He is perfect, boundless, vest, interminable, immeasurable, limitless, unlimited, unbounded, and all the other exaggerated adjectives of greatness that we can come up with. Infinity is a term the human mind cannot fully comprehend.

8 - God the Absolute is Eternal. Synonyms to help describe eternal are everlasting, undying, unending, perpetual, endless, ceaseless, timeless, interminable, without beginning or end, always existing, indestructible. God the Absolute has no beginning and no ending He has always existed and is causeless. In the world of relativity there is never a thing without a cause, because the Law of Cause and Effect is in operation, but all these causes and effects are within and of God the Absolute. There is no cause *outside* the Absolute to affect it; there is no such thing as *outside* of God the Absolute. Cause and Effect started at the beginning of time, which began with the Big Bang. The Absolute had no beginning and is causeless and He created the Big Bang. Time is relative whereas eternity is an absolute term that the human mind has trouble comprehending.

9 - God the Absolute is Indivisible. The Absolute is complete and cannot be divided into parts; it is total, comprehensive, and the ultimate thing. The Absolute is whole and there cannot be any partition, splitting up, or separation in reality. Human beings cannot see the absolute whole, so that in the relative reality of the human mind, we have to break things into parts in order to understand. We are unable to see the thing in its entirety, so we have to break it up into parts or portions of the Absolute whole. The limitation is within us since our finite mind lacks the field of view to see the whole. The human mind breaks up the whole into parts for its own convenience of understanding but in true reality the Absolute remains whole and unchanged. Every so-called part of the Absolute is always in touch with every other part of the whole; all is one, undivided and incapable of partition and separation.

10 - God the Absolute is unchangeable, constant, and permanent. God the Absolute is complete and perfect and cannot be improved. If it were to change then it would mean that it was not perfect and absolute. There is an unvarying stability and constancy about the Absolute. There cannot be any evolution, or development, or growth on its part since the Absolute is already perfect and complete. This being the case, we must realize that all that we call change, growth, improvement, progress, evolution, and development are relative terms that apply to our relative reality and not to the Absolute Reality. They are only appearances of reality to our finite and limited mind, which sees only a small-distorted part of the whole. At our level of development our mind always mistakes the relative appearance for the reality. The Absolute is outside of the Law of Cause and Effect, which affects relative things. Universal Laws are creations and tools of the Absolute serving His Divine purpose.

11 – That which is not absolute must be relative to the Absolute, or else nothing at all. All things must be either absolute or relative things and all things are aspect of the Whole. If the thing is neither of these two things then it is an illusory lie or a mistaken nothing. To humans relative things are real things; to the Absolute relative things are illusions created to serve a Divine purpose.

Chapter 2

The Human Composition

The human being is a complete system designed to provide a means for development of the Consciousness. This Human System is one unbroken system which is subdivided into ten component parts for ease of explanation, and not because they are separate and independent parts. On the contrary, all the parts are interactive and interdependent and all are necessary for the system to function properly. The ten component parts are listed in descending hierarchal order.

    10. Spirit
    9. Soul
    8. Consciousness
    7. Intuitive Mind
    6. Intellectual Mind
    5. Personality
    4. Human Instinctive Operating Mind (HIOM)
    3. Vital Energy
    2. Astral Body
    1. Physical Body

The Physical Body, Astral Body, Vital Energy, and HIOM (components 1 through 4) are components of the humans, but animals also have those same basic components. The Personality, Intellectual Mind and Intuitive Mind (Components 5, 6 and 7) are components that belong to humans only; however, it can be said that some of the higher mammals have rudimentary amounts of personality and intellect as well. Intuitive Mind (Component 7) is strictly for human use. Consciousness, Soul, and Spirit (Components 8, 9 and 10) are the Real Self of each human. The Soul and Spirit are absolute entities and do not change. The Human Consciousness however, is relative and changes as it advances forward on its evolutionary journey. Components 1 through 7 are not the Real Self; they are the "not I" components; they are merely tools that are used by the Consciousness for its own development. Following is a brief description of each of the human components.

2.1 The Human Physical Body

The Human Body is a marvel of complexity, efficiency, and beauty. It is the lowest in significance of the ten human components, and it is the only component that we know well. In the past one hundred years, medical science has made significant discoveries of the workings of every part of the human body, and in more recent years the pace of discovery has accelerated.

The basic building blocks of the human body are the cells. Cells, of course, are the building block of all living organisms on earth including plants and animals. Cells vary in size, shape, and function, but they all have the same basic structure and chemical composition. A cell is a complete miniature life that takes in food, water, and oxygen as nutrition and gets rid of waste material. It grows and reproduces by splitting itself into two identical cells, and in time it grows old and dies. The cell contains a nucleus, which contains the DNA (deoxyribonucleic acid). DNA is the chemical substance or molecule that contains the genetic program that contains the master plan for creating the bodies of living organisms. For instance, a tree cell has all the information necessary to create a tree; a zebra cell has all the information to create a zebra; a human cell has all the information to create a human. In humans, DNA information determines the color of hair, color of skin, color of eyes, the tallness of the person, and thousands of other physical features.

Cells are tiny living chemical factories that contain power plants for generating energy for their own functioning, and also contain the means for producing products and services. Some of the products that a cell produces for instance are complex proteins and hormones. Proteins are used for building other cells, and hormones are used as messengers to send information throughout the body. The cell also produces waste materials such as carbon dioxide, which it discharges into the blood stream for elimination. Cell operation is not haphazard and without direction; everything in the universe has mind and the tiny cell is no exception. Mind, which is non-physical, controls the complex operation of the cell, which is physical.

2.1.1 The Large Human Brain

A common misconception is that the human brain is the largest of the animal kingdom because the human is the smartest of all creatures on earth and therefore has a larger mind. The assumption is that the mind is hardwired into the brain, and that it occupies a physical part of the brain, but that is not quite so. The human brain is largest of all animal brains because the human brain does more physical activity; that is, it does many more varieties of physical things that other animal cannot do. It is the physical activity that requires more physical brain and not the mental activity. For instance humans can learn to ride a bicycle, drive a car, swim, read, type, play the piano, and play basketball and many more things. Humans can learn to do thousands upon thousands of physical things, which animals cannot do. The brain is not for thinking; it is for doing. The Intellectual Mind is for thinking. Another important physical activity of the brain is speech. Speech is a very complex ability and requires large amounts of brain capacity. The mind only directs these physical activities through an interface mechanism that converts the mental signals of the mind to physical signals of the brain. The mind works in conjunction with the brain, and both require a certain amount of training through repetition. A segment of the mind and a segment of the brain have to learn to do an activity through many repetitions, and once they learn the particular action, they both become proficient enough so that the action becomes automatic.

2.1.2 Computer Scientists Speculation

There is some speculation or beliefs among some modern day computer scientists that the mind is hardwired or burned into the brain, and that someday science will be able to decipher that code and copy the code into a computer of the future, and thereby have a computer that can emulate a human being. This scenario is evident in the book by Ray Kurzweil, *"The Age of Spiritual Machines: When Computers Exceed Human Intelligence"*. It assumes that the mental code can be translated into a computer language and then manipulated by intelligent computers to act like a human or even be a mental clone of the human. I disagree with the scenario because the mind is not physical and cannot be copied by physical means. Computers already have some of the attributes of the Human Instinctive Operating Mind (HIOM) since computers are good at doing repetitive things as is the HIOM. Computers can also be made to resemble or emulate human intelligence using Artificial Intelligence (AI) but that is nothing like the Consciousness. More evidence of this will be produced as we progress through the book.

2.1.3 Genetics and Cloning

Each cell of the human body is a microcosm of the entire body. Genetic scientists and engineers can now recreate a clone of a cow, or rat, or goat using stem cell from the donor animal. This should be just as possible to do with a human, but it is important to realize that the clone of the human will look physically identical to the donor, but the donor and the clone will be two entirely different people. This is the same principle as identical twins in which the twins are identical in physical looks but are indeed two different people. The DNA code contains only the physical information of the donor; it does not contain the Mind, Consciousness, Soul, and Spirit information of the donor. For instance, when the cloned physical body of the baby is born it receives a new mind and Consciousness, which will be entirely different from the donor parent. The DNA does not have the information to clone personality or intelligence of a person. Personality, Intelligence, desire, and will are part of the Consciousness and not a part of the DNA.

2.1.4 Vital Organ Subsystems

There are many specific types of cells that group themselves to form organs that perform different specialized functions in the body. These organ subsystems operate as a single unit (they form a complete whole) but we break them into separate units for the sake of understanding. The vital organ subsystems that make up the physical body system are merely listed here.

- The muscular subsystem
- The skeletal subsystem
- The skin subsystem
- The digestive and assimilation subsystem
- The respiratory subsystem
- The circulatory subsystem
- The urinary subsystem
- The reproductive subsystem
- The endocrine subsystem
- The nervous subsystem

The body is a tool of the Consciousness; it is given to us to use for our own development and evolution. It must be kept in tiptop shape (as much as practicable, and to the best of our abilities) so that it can maximize its service for us. The principles for keeping fit bodies are well known. Generally the following is suggested.

- Eat good nutritious food, but not in excess.
- Drink plenty of clean water, and breathe good clean fresh air.
- Develop a good exercise program.

- Stay busy and work hard, yet get plenty of rest and avoid stress. Too much idleness is not recommended.

The human body has not changed much in the last 20,000 years or more. It has remained essentially the same over many years except for getting a little bigger, a little stronger, and a little faster. In essence, the physical body has fundamentally stopped evolving; it is already approaching its highest evolutionary state for its required function. On the other hand, the Consciousness has developed and evolved tremendously over the same 20,000 years.

Everyone knows that the physical body is not permanent. The normal way of nature is for the body to serve its purpose and eventually grow old and die. After death, the body starts to disintegrate immediately. All the chemical elements and molecules of the body break apart into their independent parts and scatter themselves throughout the vicinity. The body is 100% biodegradable, and nothing goes to waste. Eventually these organic materials become part of other living organisms. While the physical body is temporary, the Consciousness is forever.

2.2 Astral Body

The astral body is the next higher part on the hierarchical scale of human composition. Just as the physical body is a part of the Physical Domain, the astral body is a part of the Astral Domain. The domains of the universe are discussed in much greater detail in Chapter 7 and 8, but for now a brief description will suffice. This book defines the universe as being composed of five domains as follows: 1.) Physical Domain, 2.) Astral Domain, 3.) Spiritual Domain 1 (SD1), 4.) Spiritual Domain 2 (SD2), and 5.) Spiritual Domain 3 (SD3). The Astral Domain is a parallel or complementary domain to the Physical Domain. The two domains work in conjunction and in harmony with each other. They are similar in many ways, yet are very different and perform different functions in the scheme of things in the universe. The two domains are similar in that, for everything in the Physical Domain there is a corresponding thing in the Astral Domain. For example, for every physical tree there is an astral tree; for every physical cow there is an astral cow; for every physical house there is an astral house; for every physical human body there is an astral human body, and so on. The two domains exist superimposed, complimenting each other but without interfering. The reason this works is that they are made of different wavelengths that coexist in the same space. The frequencies of the Astral Domain are much higher than the Physical Domain. Our present senses were designed to perceive the Physical Domain only; consequently we are not aware of the Astral Domain. Chapter 3 is devoted to an explanation of wavelengths and vibrating frequency. This in necessary because the entire universe including mind, energy and matter are all made of waves.

Physicists, cosmologists and astronomers are aware that there is a whole lot of matter that is unseen in the universe; they call this *dark matter*. Scientists cannot see dark matter but they know it exists because of empirical scientific data. Mathematical calculations on the total gravitational forces of the universe are not consistent with the amount of known physical matter. I do not concur with the term *dark matter* and instead calls it *astral matter*. Astral matter is of the Astral Domain and is not dark.

The physical body and the astral body coexist in a superimposed manner throughout the lifetime of the person. The physical body cannot exist without the astral body and all the physical body parts have corresponding astral body parts. For instance, for the physical arms and legs there are corresponding astral arms and legs; for the physical liver, kidney, and heart there is a corresponding astral liver, kidney, and heart; for the physical brain there is a corresponding astral brain, and so on. The phenomena of phantom limbs that amputees experience when they have lost a limb to amputation can be explained by the fact that only the physical limb is lost, while the astral limb remains. There are many reports from people who have lost an arm or a leg and continue to feel that that arm or leg is still there even though it cannot be seen. Some of these people can even experience pain in those limbs that are not there.

The existence of a corresponding astral brain creates some interesting phenomenon. This means that most sensory inputs actually have double paths through the brain. There is a sensory path to the physical brain and a simultaneous and corresponding path to the astral brain. In some cases, if a part of the physical brain is severed, the physical sense may be gone (partial sight for instance) but the person may still sense an event without actually seeing it. Subliminal sensing of an event occurs when the event is sensed by the astral sense (applies to sight, audio and smell) but not the physical sense. In actuality we have the potential to sense the Astral Domain but our conscious awareness is not allowed to perceive it yet. Those senses will become active whenever we develop our level of Consciousness to a higher degree.

The physical and astral bodies are connected to each other throughout the lifetime of the person on earth by what is called a *silver thread*. At the time of death the silver thread breaks, and the astral body separates itself from the physical body. Shortly thereafter the astral body also dies, and the two bodies begin their rapid decomposition into their discrete atoms, molecules and minerals.

2.3 Vital Energy

Energy exists everywhere in the universe; it is prevalent in every domain and in every nook and cranny of every domain. There are many variations of energies and each version has different properties and characteristics, and serves different purposes. There is physical energy, astral energy, mental energy, and Spiritual Energy. Humans of course are only familiar with physical energy, and the primary source for physical energy on earth is the sun. Physical energy includes fossil fuel (petroleum, gasoline, coal, natural gas, etc.), hydropower, wind energy, electrical energy, nuclear energy, caloric energy, etc. The physical body uses caloric energy, which we obtain from food. The astral body uses astral energy, which also comes from the food we eat, the water we drink, and the air we breathe. Caloric energy is a minuscule amount of energy uses by the body compared to astral energy. *Vital Energy* is the total energy necessary for the body to run, to motivate, to animate, to stay fit and healthy, and to stay alive. In the eastern philosophy this Vital Energy is called *Prana*. Mental and Spiritual energy are also part of the Vital Energy which humans use and are also abundant throughout the universe. Vital energy continually flows in and out of the body, and it is important to keep the body charged with as much energy as possible at all times. A body charged with lots of Vital Energy is a healthy body.

Astral energy is what scientists have called *dark energy*, primarily because they are aware of its existence but they cannot see it. Astral energy is of course not dark; it is part of the astral energy spectrum, which is made of vibrating frequencies that are much higher than those encountered in the Physical Domain. Our senses were not designed to perceive these higher frequencies. Clairvoyants however, can see astral energy in the form of human auras. They report that they can see an aura around people as well as around other warm-blooded animals. The aura around healthy people extends out from a few inches to a few feet, and is multicolored as a rainbow.

Young healthy people normally have abundant astral energy while older people are deficient and consequently prone to get sick. The health of a person is apparent in the aura. Healthy people have auras that are bright and vivid and extend way out. A sickly older person will project an aura that can barely be seen around the surface of the body and is dull in intensity. With training, a clairvoyant can identify the intensity, vigor, and health in a person by the intensity and radiance of the aura. By the same token a clairvoyant can identify the depletion of energy of a weak, or sick, or older person by the dullness of the aura. The astral atmosphere is always charged with astral energy, but it contains more energy in the daytime when the sun is out then it does at night. That explains why a sick person is likely to suffer more discomfort at night and is likely to feel better in the daytime. That also explains the statistic of why more sick people die at night than die in the daytime.

At the moment of death, all Vital Energy leaves the physical body as well as the astral body and both bodies immediately start their normal process of decomposition. The energy is then absorbed by the infinite universal energy supply and recycled again in some other life.

2.4 The Human Instinctive Operating Mind (HIOM)

Instinctive mind was the first mind to appear in the universe. It appeared in its most primitive form at the time of the "Big Bang". Instinctive mind is the operating mind in the physical and astral domains. Its function is to operate everything on earth to include all non-living (inanimate matter) as well as all living organisms. Instinctive mind is not a thinking mind; it is a doing and organizing mind that operates in an automatic and subconscious manner. It is good at doing repetitive things. Instinctive mind has evolved from the most primitive to the most advanced form in humans today. The evolution of instinctive mind starts with the Mineral Instinctive Operating Mind (MIOM), which operates all the micro non-living materials of the universe. It provides all the operating instructions for physical and astral matter and energy. MIOM primarily provides all the operating methodology for matter and energy at the atomic and molecular levels, and is the foundation for the Laws of Physics as we have been able to interpret them. The other instinctive minds that operate the non-organic functions of the Physical and Astral Domains are Terrestrial Instinctive Operating Mind (TIOM) and Galactic Instinctive Operating Mind (GIOM). The next higher levels of instinctive mind, on the scale of evolution, are the ones that operated the functions of living organisms starting with Plant Instinctive Operating mind (PIOM), which operates the functions of all living plants from the most primitive to the most advanced. Next on the scale is the Animal Instinctive Operating Mind (AIOM), which operates all the function of animals from the most primitive to the most advanced mammals. The Human Instinctive Operating Mind (HIOM) is the most advanced of all instinctive minds.

The Human Instinctive Operating Mind (HIOM) is the mind that operates and regulates every part and function of the human body in an automatic mode. HIOM operations in a subconscious manner since the functions are beyond our level of awareness. It works in conjunction with the brain and autonomic nervous subsystem, which operates all the vital organs of the body. For instance, the HIOM operates the heartbeat in the circulatory subsystem, the breathing in the respiratory subsystem, the removal of toxins in the urinary subsystem, the digestion and assimilation of food in the digestive subsystem, and performs thousands of other services. In other words, HIOM keeps us alive.

Much of the HIOM can be thought of as being composed of modules like computer software modules, with each module performing a specific function. For instance, there is an HIOM module that operates the liver, there is an HIOM module that operates the kidneys, there is an HIOM module that operates the heart, there is an HIOM module that operates the lungs, and so on. The biggest part of the HIMO however, operates out of the brain. Output commands from the brain to the discrete modules travel through the nervous system and input information from the sensory organs to the brain also travels through the nervous system. Most of the HIOM is superimposed on the physical brain and in the space immediately surrounding the brain, but discrete modules of HIOM are distributed throughout the body; they are usually located in the vicinity of the vital organ that it serves.

2.4.1 Computer Analogy

The HIOM can be compared to the operating system (OS) of the modern computer. The OS in the computer is the software that organizes the operations of the computer's hardware and application programs. The computer OS performs many functions that are hardly noticeable to the user, and can be compared favorably to the functions of the HIOM that are performed in a subconscious mode. In the computer the programs are stored in the hard drive. When the program is called upon for use, it is downloaded to the random access memory (RAM). Similarly, the HIOM module that is to be used is moved to the proximity in the brain area where it is to be used. The brain is equivalent to the computer's CPU and RAM in this case. The human does not have a physical hard drive for storage of the HIOM. Most of the HIOM is simply suspended in the space immediately surrounding the physical brain, but much of it is also distributed through the rest of the body.

2.4.2 The Trainable Mind

The HIOM is composed of the summation of MIOM, PIOM and the AIOM. The functions of the MIOM are still present in the human body since we are composed of atoms and molecules that the MIOM operates. Likewise the functions of the PIOM are still present in the human body since we are made of living cells that the PIOM operates. Similarly the functions of the AIOM are still present in the human since AIOM operates the vital organs of animas as well as humans. Not only is the HIOM the summation of all the instinctive operating minds but in addition to that it also has a "Trainable Mind" which gives the human a tremendous advantage over the other mammals. The Trainable Mind is almost nonexistent in animals but is extremely influential in humans.

The Trainable Mind is what helps us learn new activities, which become almost automatic once the activity is learned. The Trainable Mind learns to do such things as ride a bicycle, learns to swim, learns to play pool, learns to play basketball, learns to play the piano, learns to type, and most important it learns to speak. The Trainable Mind is not a thinking mind; it is learning and doing mind that learns by repetition. Initially the Trainable Mind is under the direction and supervision of the Intellectual Mind. Once the Trainable Mind learns to do something well through repetition, it generally takes over control of the function allowing the Intellectual Mind to do other things. The Trainable Mind takes over the drudgery of repetitive activities from the Intellectual Mind.

A scenario to illustrate the initial learning process for the Trainable Mind and the brain is as follows. Think back to your youth when you were trying to learn to ride the bicycle for the first time. In the initial learning process the Intellectual Mind has to provide commands for every detail of the riding. For instance, the Intellectual Mind has to command the left foot to press down on the left peddle, and then the right foot to press down on the right peddle, then the left and then the right. At the same time you have to try to stay balanced and stay upright, and also at the same time, control the steering and go straight but making sure to go around the tree. The Intellectual Mind cannot continue to provide detail commands for the Trainable Mind for long. Detail commanding is tedious work that the Intellectual Mind was not designed for. Luckily, with practice the Trainable Mind can learn to perform the process automatically and eventually relieves the Intellectual Mind of the drudgery. With time the Trainable Mind becomes proficient at riding the bicycle and eventually it takes complete control over of the process. Riding the bicycle becomes an effortless subconscious operation. At that stage the process becomes a part of the HIOM and the Intellectual Mind becomes free. Animals do not have a sophisticated Trainable Mind; therefore, they can only learn a few new tricks at a time. Dogs, for instance, can learn to fetch, sit, roll and a few other things, which are cute, but somewhat un-natural.

Cats naturally learned to stalk its prey and to pounce on it and kill it for food. The cat is a master predator and efficient killer, which it has to do well in order to survive in the wild. But the cat is not efficient at learning any more new things like the human can. The cat therefore does not require a large brain like the human.

2.4.3 HIOM and Brain Interaction

The none-physical HIOM does most of its work through the human physical brain. There is a common misconception that the mind takes up space in the physical brain as if it were hardwired into the brain, but that is not so. The HIOM is superimposed and suspended in the space immediately surrounding the brain, but it is not a program that is burned into the brain. It is not like the Read Only Memory (ROM) or the hard disk drive of the computer. The brain is a command and control central for the HIOM, and the bulk of the HIOM operations through the brain but it is not hardwired in the brain. A large amount of the HIOM is also distributed throughout the body, and operates many automatic functions of the body without going through the brain.

The HIOM works through the human brain by converting non-physical mental signals to physical brain signals through a special interface mechanism. All body motions are initiated mentally by the HIOM and translated to the brain for the actual physical motion. Speech for instance is very complex and is a part of the function of the Trainable Mind. Speech is composed in the Trainable Mind in the form of a few words at a time and then the HIOM triggers the area of the brain which has learned to say those words one word at a time. To say words, the mind and brain work in conjunction and have to simultaneously coordinate the muscles of the lips, the chicks, the tongue, the throat, the lungs, the breathing, and many more muscles. Speech requires a large amount of brain capacity and it is because of this larger brain that humans are able to talk. Animals cannot talk because they do not have the brain size that is necessary to carry on the operation. Animals have to communication using simpler methods such as grunts, or squeals or other body language. Most of the process of speaking is carried out in a more or less subconscious manner. Humans need to communicate through advanced forms of speech in order to continue our growth at our present level of evolution. Animals, conversely, do not need an advanced form of communication for their level of evolution; at their level grunts, squeals, howls and such forms of interaction are good enough. The large brain, the Trainable Mind, and speech are tools that have been granted to humans so we can continue to advance.

2.5 Personality

Human Personality is the fifth component part in the hierarchy of the human composition. The first four component parts that have been discussed thus far are all shared with the animals, but Personality is the first of the truly human traits that are not shared with the animals. That is, both humans and animals have physical bodies, both humans and animals have astral bodies, both human and animals have vital energy, and both humans and animals have instinctive minds. All higher mammals, especially the primates such as the gorillas, chimpanzee, and orangutan have personality traits, but they are rudimentary compared to the complex personality of the human and they are not considered to be real bona fide personalities.

Human personality is complex and is formed by many factors, and most of these factors lie in the subconscious mind. Consequently, for the most part, we are unaware of why we are the way we are, and why we act the way we act. Five of the main factors that affect our personality are:
1. Childhood experiential environment – subconscious
2. Lower emotions from the HIOM – subconscious
3. Past life experiences – subconscious
4. Soul-Consciousness contributions – subconscious
5. Molding by the present personal environment – conscious

A brief explanation of each factor follows:

2.5.1 Childhood Environment

Environment at birth and the early childhood experiences are important. A child that starts life in a stable home with love, joy and peace has a good chance for shaping the personality in a good way. Parents or guardians that provide the essentials for early life such as food, shelter, and clothing, as well as love, guidance, and discipline will contribute to a child's stable personality. On the other hand, an unstable home environment with violence, abuse, poverty, drugs, and alcohol, will produce deep scars in the personality of the child, which will handicap the child throughout that person's lifetime. The inputs that form this part of the personality are considered to remain in the subconscious mind since most people cannot remember the event that took place in their infancy and early childhood. The subconscious mind is like a database of the Consciousness of each and every individual person. This database stores everything that the person experiences throughout his or her lifetime. It is a huge database that has no limit, it is permanent, and it never forgets. Negative data that affects the daily life of a person is in the foreground of the subconscious and needs to be sent to the background. It is important that a person become aware of those negative hindrances and by force of will replace negative traits with positive traits. Negative traits will, with time, fade to the background of the subconscious when they are replaced with positive traits.

2.5.2 Lower Emotions from the HIOM

It was mentioned in the previous section that the AIOM is part of the HIOM. The AIOM contains all the information to run all the animal's physical body functions (the entire vital organ functions for instance), and likewise as a part of the HIOM it also runs all the human's physical body functions, which is good. However, the AIOM also contains the lower negative emotions of the animal, which is bad garbage. These negative emotions were necessary for the survival of the animals at the various stages of their evolution, and since the HIOM is a progression from the animal instincts, we inherited these lower emotions from the animals. The HIOM is a wonderful mind, but it also has a lot of trash in it. We inherited both the good and the bad. These animalistic emotions have a pronounced influence on our present personality. The less advanced people on the evolutionary scale are more likely to be influenced by the lower emotions, and cause them more havoc. On the other hand, the more advanced people on the scale of evolution are likely to suffer less from the lower emotions. The reason for this is that they have advanced sufficiently to learn to moderate the emotions so they are less harsh. These emotions are a part of our makeup, and can never be completely erased. However, with conscious effort, these negative emotions can be altered and softened by replacing them with equivalent positive emotion. It is important that a person learns to suppress the negative emotions and to accentuate the positive emotions. This can be done gradually by force of will.

2.5.3 Past Life Experiences

Our present personality is shaped to a large extent by the experiences of our past lives. Remember that the Consciousness stores all experiences to include the present life's experiences as well all past lives experiences. The experiences of the most recent past lives have more impact on us than past lives that are far removed. What we call phobias in the present life are actually destructive experiences of a previous life, which remain in our subconscious mind as part of that permanent database. For instance, the fear of water may be due to death by drowning, the fear of fire may be due to bad experience or death or mutilation by fire, claustrophobia may be due to death by suffocation, fear of heights may be due to death or mutilation due to falling, and so on. Good personality traits can be due to idyllic experienced of a past life. Happiness can be the result of the joy of raising a family, or the joy of music, or the joy of building things, or the joy of higher education, or the joy of religious pursuits, or the joy of travel, and so on.

Susceptibility to criminal behavior is often due to a life of criminal experiences in the past life. For instance, a person, who by chance participates in a war in a previous lifetime and committed atrocities as a matter of course, would probably be a person that can kill in this lifetime without remorse. A deprived person who made a living by begging or stealing in a previous life would probably think nothing of stealing in this life. An embezzler in a previous life would probably be a swindler or conman thief in this life. These negative traits however are subject to the law of Cause and Effect and the person will sooner or later have to pay for those crimes. This law is always instrumental in causing the criminal to become aware of the errors and eventually repent and go straight. It is all part of our evolutionary process.

In most cases, homosexuality is due to the influence of sex change form one lifetime to the next. For example, a person that had a happy and well balanced life as a female in the previous life and then happens to be born a male in this life will most likely have a strong desirability for males in this life. This is so because the old needs and desires do not evaporate from one lifetime to the next. The change in sex from one lifetime to the next is fairly common, and it does not happen by chance or accident. A sex change normally happens because the Soul-Consciousness feels that this kind of change is necessary to balance the evolutionary experiences of the Consciousness.

2.5.4 Soul-Consciousness Contributions to the Personality

The Soul and the Consciousness plays a important part in the personality of a person. They are both instrumental in selecting the baby's parents, race, religion, and environment prier to the baby's birth. The selections are made by the Soul and Consciousness in accordance with the requirements of the Consciousness for further growth and development. Often there are conflicts between the Soul and the Consciousness in the selection process. The conflicts arise due to the somewhat different priorities of each. The Consciousness tends to prefer rebirth into the same environment of the previous life; this is because of the comfort of familiarity. The problem with that is that there may not be much potential for growth and development since it is going back to relive the same old life. The Soul, on the other hand, will stress a different environment that will offer different experiences and more possibilities for positive development. The conflict is always resolved amicably, but the selection of parents and environment by the Soul-Consciousness remains hidden from the conscious mind. The conscious mind is called the *Intellectual Mind* and is the subject of discussion for the next section.

2.5.5 Molding by the Present Personal Environment

The part of the personality that you are aware of is formed as a result of the environment that you live in. This is the part of the personality that makes you think you know who you are. If you were asked, "Who are you?" you would probably respond by describing yourself and saying, "My name is 'so and so'.", "I live in 'such and such a place'.", "I am 'so tall'.", "I have 'so much education'.", "I like to travel to 'such and such places'.", "My hobby is 'such and such'." "My religion is 'such and such'." "My profession is 'such and such'". What you are doing here is describing yourself, as you know yourself. That is the response of most people, but that is the description of the physical self. The Personality is temporary and it

Evolution of Consciousness

is not the Real Self. The Real Self is the Soul-Consciousness, which is the eternal you.

2.5.6 The Changeable Personality

The personality is very changeable. It changes throughout the lifetime of the person. It can change in small incremental steps through small experiences of everyday life, or it could change drastically with dramatic changes in life. A dramatic change may be the loss of a parent when the kids are young, which can cause a change in lifestyle and consequently a change in personality. Getting married and having children with increasing responsibility for home and family always affects the personality. The horrible experiences of war or violent crime can also dramatically affect the personality. More gradual changes come with completion of educational milestones such as high school and then college and also with starting a new career. This usually has positive affects on the personality.

Almost all people that are over fifty years or older can look back at their present life and agree that their personality has changed a great deal over the years. They can point out that since they were twenty years old to the present time, they have changed tremendously. Some of these older folks may declare "I am not the same person now, as I was at twenty." or they may say, "Looking back to when I was twenty years old, I can truly say I was pretty dumb then". Or they might say, "Sometimes I wonder how I even survived my early years". Or you might hear them say, "My experiences over the years have changed me very much; I feel more mature and stable now." Another common saying is "I wish I knew then what I know now." These kinds of expressions are common among older folks. These changes are signs of positive "Evolution of Consciousness" over the present lifetime.

2.5.7 Heredity

A common misconception is that personality and intelligence is inherited genetically from the parents, but this is not so. Genetics (DNA) only transfers the parent's physical characteristics whereas personality is mental, and mental traits are not inherited through genetics. The reason that the personalities of the parents and the offspring often seem to be similar is that the Soul-Consciousness selects the parent before birth. The Soul-Consciousness tends to select parents that are compatible and have similar characteristics as themselves. This is an application of the universal "Law of Attractions". Parents also teach their own character traits to their child while the child is growing up. Many of these learned character traits often become part of the personality.

2.6 The Intellectual Mind

The *Intellectual Mind* is the sixth component in the hierarchy of the human composition. It is a truly human component that is not shared with the animal kingdom. The Intellectual Mind is our conscious mind that gives us our awareness of being. It is the mind that gives us the power to think, to reason, to analyze, to scrutinize, to make determinations, to make interpretations, to make decisions, and to be conscious of being human. Animals do not have this mind. The more advanced mammals and particularly the primates have traces of Intellectual Mind and appear to have some rudimentary power of reasoning, but it is very primitive and not classified as a true Intellectual Mind.

A popular misconception is that the mind is stored inside the brain; similar to the way that computer software is stored in a computer's hard drive. The mind is not physical; it is made of high frequency vibrations that are outside the realm of the Physical Domain. The Intellectual Mind resides in the immediate vicinity of the space surrounding the brain, but it is not hardwired into the brain and it does not interface directly to the brain. The Intellectual Mind has a direct interface to the HIOM, and the HIOM then interfaces to the brain. All physical work is actually done by the HIOM working through the physical brain. The Intellectual Mind is used for thinking and reasoning while the HIOM is used for doing work. When the Intellectual Mind needs something done it commands the HIOM to do it and the HIOM obeys.

2.6.1 Chain of Command

The Intellectual Mind is master over the HIOM as well as the Trainable Mind. The HIOM and the Trainable Mind are not thinking minds; they are doing minds. Much of the HIOM works in an independent and automatic fashion but parts of it are under the direct control of the Intellectual Mind. The Trainable Mind is under the control of the Intellectual Mind while it is learning and becoming proficient at performing an activity. Once the Trainable Mind learns to do an activity well, the Intellectual Mind releases it to act in a more or less independent way.

The chain of command for physical activity actually starts with the Consciousness. The Consciousness is the Real Self and in advanced humans it is the initiator of all activity. In advanced humans the Intellectual Mind provides the bulk of the direction; whereas, in the less advanced humans the instinct tends be more in control. The chain of command to do a physical activity follows from top to bottom:

1. Consciousness – The Consciousness initiates the activity by commanding the Intellectual Mind to do something. This command is always brief and with little detail. For instance, the command may be equivalent to "Go to the store and get milk."
2. Intellectual Mind – The intellectual Mind receives the command, analysis it, and fills in all the other details to carry out the command. It knows what to do because of previous knowledge and experiences. The Intellectual Mind then issues many small and brief commands to the HIOM. There is a brief command for every main activity needed to do the job. For instance, there may be a series of commands such as "Look for the car keys." once the car keys are found the next command may be, "Walk to the car." once the person gets to the car the next command may be, "Get in the car, start the car, and drive to the store."
3. HIOM – The HIOM takes each command from the Intellectual Mind and adds greater detail for every motion required. Remember that the HIOM is not a thinking mind, it is a do-

ing mind. The HIOM commands the brain in every minute detail. All activity between the HIOM and the brain is carried out in a subconscious manner. The interface between the HIOM and the brain is done by an interface mechanism that converts nonphysical mental signals to physical brain signals.

4. Brain – The brain receives detailed information from the HIOM and adds even greater detail to those instructions. Like the HIOM, the brain is not a thinking instrument, it is a doing instrument. It learns to do things through repetition much like the Trainable Mind and once the action is learned it becomes hardwired (figuratively) in the brain. The HIOM merely triggers that tiny section of the brain that has learned that particular action. The neurons in the brain generate electrochemical and biochemical signals that sever as the transmitters of information through the nerves.

5. Nerves – Next, the nerves distribute the electrochemical signals that are produced in the brain, to all the individual muscles. The signals provide detailed, coordinated, and simultaneous instruction to multiple muscles where the physical movement takes place.

6. Muscles – Groups of muscles react to the electrochemical signals in such a way as to produce motion that is smooth and well coordinated. As you can see the entire process is enormously complex, and most people are oblivious to these wonders.

Messages from the outside world to the mind are similar but in reverse order. The normal path for the sensory perception of sight for instance is as follows. Light enters the eye and activates the optical nerve system, which activates proper neurons (cells) in the brain. The neurons then translate the sight message from the physical to the mental and the message is sent to the HIOM. The HIOM then sends the message to the Intellectual Mind where it is interpreted and analyzed and a decision is made as to what to do next. The Intellectual Mind informs the Consciousness of the activity and the Consciousness has the option to allow the Intellectual Mind to continue with the process or it may intervene and alter the Intellectual Mind's activities if it desires.

There are provisions in the structure to accommodate emergency reactions through the nervous system. An emergency reaction such as removing the hand from a hot plate takes place in the HIOM without sending the message to the Intellectual Mind until after the fact. This situation is for the purpose of reducing the reaction time to prevent further damage due to an accident. This reaction takes place within the arm, hand and the HIOM in the arm and hand. This message does not go immediately to the Intellectual Mind for interpretation and analysis. There is no time for that when a serious burn is imminent. The message goes to the Intellectual Mind after the fact, that is, after the hand has been removed from the hot plate. The Intellectual Mind then has the option to analyze and make decisions.

The Intellectual Mind is master of all components below it in the human hierarchy but it is subservient to the Consciousness, which is above it. The Consciousness is the Real Self and has mastery and authority over the Intellectual Mind. If the Intellectual Mind is not disciplined to exercise its authority over the HIOM it can loose that authority, and become the slave to the HIOM instead of its master. Of course that is the way the animal operates. If the Trainable Mind is trained with bad habits, then those bad habits will tend to dominate the actions of that person. Bad habits will eventually have to be corrected and suppressed.

2.7 Intuitive Mind

The *Intuitive Mind* is the seventh component in the hierarchy of the human composition, and is classified as a higher subconscious mind. For instance, the Intellectual Mind is a conscious mind and fits between the lower subconscious mind which is the HIOM and the higher subconscious mind which is the Intuitive Mind. The Intellectual Mind is aware of itself but normally not aware of the HIOM or the Intuitive Mind on either side. The Intellectual Mind can easily accept the two subconscious minds by simply becoming aware of their existence. It is not difficult for the Intellectual Mind to accept intellectually the two subconscious minds on either side of it.

The Intuitive Mind is available to all people to use regardless of the level of their personal growth. In practice the less advanced people are not aware of it and do not use it except accidentally at times. The more advanced people of the Human Consciousness Stage use it sparingly. People in the higher stages of evolution such as the people at the Soul Consciousness Stage use it to a much greater extent. It behooves people at any stage to try to learn to use it because of its tremendous potential benefit. In order for a person to benefit from the Intuitive Mind, that person has to learn to acknowledge its existence and make a conscious effort to use it.

Many of the good ideas and the finer things in life come from the Intuitive Mind. Things such as the love of the arts, music, beauty, poetry, philosophy, theology, metaphysics, science, and mathematics, all come from this subconscious region of the mind. These things are not reasoned by the intellect, their source is the subconscious region, the Intuitive Mind. The intellect is cold and the Intuitive Mind is warm with high feelings and emotions. The Intuitive Mind does not run contrary to the intellect; it simply goes beyond the intellect. The Intuitive Mind passes down truths, which cannot originate with the intellect. The origin of thoughts about God, and spiritual insights come from the Intuitive Mind, and not from the intellect. Once these kinds of thoughts and impressions are transferred to the Intellectual Mind, the intellect can reason on them and they become part of the conscious thinking. The intellect normally has no problem trusting the inputs from the Intuitive Mind. The Intuitive Mind is not a thinking mind it is a knowing mind.

2.7.1 Brainwashing

The intellect may sometimes be brainwashed or influenced into resisting the inputs from the Intuitive Mind. This brainwashing can come from peer pressure, educational training, philosophical dogma, or religious beliefs. It can be a real problem for some people because it can hold them from utilizing this great source of goodness in nature. Some people can overcome the effects of brainwashing fairly well, while others may take years, and yet others may not overcome it during the present lifetime.

2.7.2 Access to Higher Realms

The Intuitive Mind can be used to provide access to the higher realms of the universe. Access to the Akashic Records can be obtained through the Intuitive Mind. The Akashic Records is the huge database of the Universal Mind; it is the universal memory of nature whose information is infinite. The Intuitive Mind can also provide help from the Spiritual Guides, Spiritual Teachers, Spiritual Counselors, Angels, and Archangels. Some people establish this path through the technique of prayer or meditation since both can provide similar results; it is up to the individual preference. Some writers refer to the Intuitive Mind as the Spiritual Mind or the Christ Consciousness. Both terms have the same basic meaning. People of various religious denominations feel that Christ Consciousness gives them direct access to God.

2.7.3 Genius

Genius comes from the Intuitive Mind also. Musicians such as Amadeus Mozart, writers such as Mark Twin, inventors such as Thomas Edison, scientists such as Albert Einstein, and industrialists such as Henry Ford, got their inspiration from the Intuitive Mind. Every one of these geniuses, most likely, has a different technique for gaining access to their information. Some people get their information through prayer, others through intense contemplation and focusing on a subject, others through meditation. All of these techniques are good, and every one is entitled to there own preference. Mozart once wrote: "I cannot really say that I can account for my compositions. My ideas flow and I cannot say whence or how they come. I do not hear in my imagination the parts successively, but I hear them, as it were, all at once. The rest is merely an attempt to reproduce what I have heard." Burt Bacharack composed many songs during the 1960s and 70s. Once he said in an interview that many times when he was having trouble with a musical composition he would go to the bathroom and while sitting on the toilet he would continue to go over the song in his mind. By the time he was done he would have completed the composition in his mind and was ready to go write it down on paper.

2.7.4 Personal Experience

My personal experience as an electronics engineer in the late 1960s, 1970s, and early 1980s is that I learned to focus on a design problem. I could immerse myself mentally into the problem where the outside world seems to become blurred. In this state of concentration and focusing I could get ideas and resolve problems. As the problems were resolved in the mind, the results were recorded on paper in the form of equations, words, sketches, or schematic diagrams. A complete electronic system is the result many of these small design problems being resolved. Many of the ideas for complex designs come from past design experiences and intellectual reasoning, but the new stuff that has never been done before, has to come from the Intuitive Mind. It is of great comfort to know that the Intuitive Mind is always there when you need it. With practice the process becomes easier, and more direct. Personal self-confidence increases due to increased reliance on the Intuitive mind.

Reasoning does not necessarily have to be done within the conscious intellect. Much of the reasoning should be relegated to the Intuitive Mind, which obtains the information and passes it back to the Intellect. Once you get the hang of this you want to use it as much as possible. I have friends that take engineering problems home to think on the problem overnight, and come to work the next morning with the solution. Of course, engineers are not the only ones that can use this method of thinking. Anybody who has a complex problem to resolve can use the technique of concentration and focusing, preferably in a quiet place with little or no interruptions. Practice can make you proficient, proficiency gives you confidence, and confidence will open new doors for progress, advancement, and joy.

2.7.5 Meditation Technique

The Intuitive Mind can provide the Intellectual Mind with a lot of good information, which may be obtained either consciously or unconsciously. There are people who have by chance stumbled on ways to reach information from the Intuitive Mind, and are usually not aware of where that information is coming from. The preferred method is to reach the information consciously through meditation or prayer so that you can have direct control. There are several meditation techniques that are used to access information from the Intuitive Mind. This book recommends the "Silva Mind Control Method" or the "Silva Ultra Mind" technique by Jose Silva. Silva offers a scientific, well researched, and well thought out method that is designed to help you learn to take full advantage of the Intuitive Mind. He has several books in print, and also offers home study courses in audiocassettes or CDs. Also available are workshops that are conducted throughout the United States and abroad.

2.8 The Consciousness

Consciousness is the eighth component of the human composition. Consciousness is part of the Real Self, and like everything else in the universe it is made of waves that vibrate at extremely high frequencies. These frequencies are much, much higher than anything we can ever imagine in the Physical Domain. The Consciousness is eternal and indestructible. It is eternal, because there is no such thing as time and space in its true being. In actuality, while living an earth-life in a physical body the Consciousness does experiences time and space, but its true nature belongs to a Spiritual Domain which does not contain time and space. The fact that there is no such thing as time and space is a difficult concept for many of us to accept because our senses were not designed to perceive a world with no time and space. Our senses were designed to perceive in the Physical Domain only because that is all that is necessary for our present stage of development. Mother Nature's plan is that as we develop past our present stage, we will be given additional senses as needed to meet the new requirements. It is possible to train the intellect to accept the unseen and to accept things on faith if given enough logic and backup on the subject matter. For instance, we cannot see a round earth, yet we have learned to accept the earth as being round; we cannot see mind yet we accept it as existing because we can see the evidence that it demonstrates.

While living in a physical body in an earth-life, it helps some people to visualize the Consciousness as being real and residing in the body. You can visualize the Consciousness as being oval in shape, about 12 inches high and about 8 inches around the middle, and occupying a space in the chest cavity, actually superimposing itself in the vicinity where the heart and lungs are located. You can visualize the Consciousness as being active. The surface of the oval is reported to appear as having flashes or sparkles of light; the light of course is not of the electromagnetic variety, but of the Spiritual Domain variety. These flashes of light can be thought of as similar (for illustration only) to the sparkles of a sparkler commonly used during Fourth of July celebrations. Each person is different and each person is encouraged to develop his or her own method of visualization; whatever works for that person is good enough.

Most people in our society belong to the upper levels of the Human Consciousness Stage. At this stage in our evolution we think of ourselves as physical bodies with a brain that happens to be able to think. But, in actuality we are much more than that; we have to learn to think of ourselves as a Consciousness that has a physical body and mind. The physical body, the astral body, the HIOM, the personality, and the Intellectual Minds are temporary tools of the Consciousness to use for its own evolution. The Consciousness acquires a new body and mind for each new life on earth, and uses this new body and mind for the important purpose of its own development. The body and mind are temporary tools and therefore not the Real Self. Only the Spirit, Soul and Consciousness are the Real Self. The Consciousness cycles through many lives on earth, and with each life cycle it gains in experience, knowledge and wisdom. All these experiences, knowledge, and wisdom are cumulative and are stored in a giant database that misses nothing and is forever. The summation of all these experiences, knowledge, and wisdom contributes to the level of evolution of the Consciousness. Onward development is the primary purpose for the existence of the Consciousness. As a matter of fact, God created the whole universe for the express purpose of providing a platform for the Consciousness to evolve. The universe was not created to have moons, planets, suns, stars, galaxies, black holes, etc. it was created for the Consciousness. The Consciousness is the most important thing in the universe.

The Consciousness is a relative entity and consequently changes and evolves. Evolution of Consciousness takes place over thousands upon thousands of earth year to complete, but earth years are relative units and therefore not real in the absolute sense. The Consciousness is aware of time and space only while living an earth-life. In actuality the Consciousness exists in a Spiritual environment where there is no time and space. This principle is a duality of life, which means that the Consciousness lives earth-lives with time and space, as well as, Spiritual life with no time and space.

Consciousness is the master of all the seven human components parts that have been discussed thus far in this chapter. The lower component parts are merely tools of the Consciousness to be used by the Consciousness for its own evolutionary process. The body and mind are slaves to the Consciousness and are commanded by the Consciousness.

2.8.1 Desire and Will

An important characteristic of the Consciousness is that it has *desire and will*. Desire and will comes from the Consciousness and not from the Intellectual Mind. When the Consciousness passes a command to the Intellectual Mind, the command goes with a certain amount of desire and will. The thoughts that are created as a result of a command from the Consciousness are impregnated with mental energy and the amount of mental energy imparted to the thought is dependent on the strength of the desire of the Consciousness. The stronger the desire is the more the mental energy, and the stronger the power of the thought. This is called "will power". The amount of mental energy imparted to the desire and will varies in quantity and degree, and therefore the strength of the command also varies in degree of force and determination. A weak command (with little energy) may not even be completed, where as, a strong command (with lots of energy) can move mountains.

2.8.2 The Consciousness is Subconscious

It seems ironic, when we think about it, that the Consciousness, which is the Real Self, is in our subconscious realm, and we are not consciously aware of it. We have a conscious awareness of being human beings through our Intellectual Mind, but the Intellectual Mind is only a tool of the Consciousness, therefore it is limited in its conjecture. The Consciousness is, among other things, a huge database that contains all the details of every experience that ever occurred in all previous lives and it is all outside the range of our conscious awareness. We were designed that way. A small amount of contemplation on the matter would reveal that if we were conscious of our Consciousness we would probably be overwhelmed with its contents and we could not function. Consequently only the gist or the average of the necessary experiences becomes available to us at the proper times. It is in Mother Nature's plans to eventually let us become completely conscious of our Consciousness, but that will have to wait until we advance to our next stage of Consciousness, which is the Soul Consciousness Stage.

2.8.3 Analogy of the Automobile

An analogy used to help explain the function of the Consciousness is to think of the body and mind as equivalent to a brand new automobile. A new automobile has every thing in good working order including the engine, tires, brakes, steering mechanism, and everything that makes it ready to go, but a new car cannot go anywhere without a human to drive it and direct its every move. By the same token a perfect human body and mind is nothing without the Consciousness to direct it. A perfect body and mind cannot survive for very long without a Consciousness to animate it.

2.9 The Soul

In the hierarchy of the human composition, the Soul is the ninth human component. A term that is used to describe the Soul is that it is a "Spark of God". This term is used to indicate that it is a part of God and is made of God's substance. An analogy of this principle is that the rays of the sun are made of the sun's substance, but they are not the sun, and by the same token, the Soul is made of God's substance, but it is not God. The Soul is absolute, eternal, and indestructible. It is difficult to give a precise definition of the Soul because, being made of God's substance, it has an infinite quality which our mind cannot comprehend.

Each human being is an individual Soul and each Soul has an individual Consciousness attached to it. This Soul-Consciousness combination is the "True Self or Real Self" and this arrangement lasts for the duration of the evolution of the human Consciousness, which takes place over many thousands of earth years. The Soul is a piece of the Spirit; they are both absolute entities and do not evolve. On the other hand, the Consciousness is relative and consequently it evolves.

2.9.1 Function of the Soul

It is the function of the Soul to guide the Consciousness through the path of unfoldment and growth. As the Consciousness goes through its trials and tribulations on its journey through many earth-lives, the Soul guides and coaches the Consciousness along the path. It guides without corrupting or compromising the millions upon millions of steps needed to grow. The Soul guides yet it also allows the Consciousness to experience things on its own, to learn every lesson on its own, and to fend for itself through life. There is no way to cheat or defraud the system. The Soul knows that the Consciousness starts out ignorant and that it has to learn its lessons, one step at a time. It will not carry out a task for the Consciousness in order to save it from the nuisance of learning the experience. The Consciousness will not be allowed to skip any of the lessons or experiential processes necessary for its growth.

An analogy to help illustrate this principle is that of a wise mother who loves her child but who allows the child to struggle with a task so the child can learn to do the task. The mother knows that she can coach and help the child but she cannot learn for the child. One example that comes to mind is when a child is trying to learn to tie his or her shoelaces and the mother watches her child struggle but will not step in to help. The mother knows that she must leave the child alone to learn on his or her own. The child has to learn many things on its own by doing, by making mistakes, by correcting those mistakes, and by continuing to grow forever forward. But throughout the child's learning process the mother keeps that watchful eye of protection and is ready to step in with the motherly helping hand if necessary.

In another mother and child analogy, suppose that the child wants to go out and play with his or her friends and asks the mother "Mom, may I go play with Johnny?" and the mother replies "Yes, you may go play with Johnny but not until you finish your homework." In a similar method the Soul guides the Consciousness on its way to growth and advancement.

2.10 The Spirit

In the hierarchy of the human composition, the Human Spirit is the tenth and highest component. The Spirit like the Soul is difficult for us to define because it is a manifestation of God the Absolute, which is infinite in its nature and we cannot define infinity. It is made of God's substance and contains many of God's qualities and attributes. Since it is difficult to give a good solid definition because there are no worldly experiences to refer to, we have to resort to crude analogies that will provide a basic idea.

2.10.1 Analogy of the Sun Shining on Dewdrops

Suppose that there is heavy dew on a field of grass in the early morning, and as the sun rises over the horizon you see the reflection of the sun on millions of tiny little dewdrops. Those reflections look like millions of tiny suns on the grass. The reflections are not an illusion; they are made of real substance of the sun, yet they are not the sun. The real sun is still up in the sky.

This is an example of the Divine paradox, which states that, *"The One is in the many, and the many are in the One"*. The sun may shine on millions of dewdrops, and the droplets may reflect millions of tiny suns. Yet while each droplet contains a bit of the sun, there is only one sun, which remains in the sky. The Human Spirit is analogues to the reflection of the sun on the dewdrop; it contains a tiny bit of the ingredients, qualities, and attributes of God. The Spirit is a "drop from the Sprit Ocean; a grain of sand from the shores of the infinite, a particle from the sacred Flame." Yet, Human Spirit is not God the Absolute; it is only an emanation of God the Absolute.

2.10.2 Analogy of a drop of Water from the Ocean

Another crude illustration that is appropriate states "The Human Spirit is to God, as a drop of ocean water is to the ocean". The significance of this proportionality is that the drop of water from the ocean has the same chemical composition as the ocean, yet it is not the ocean. By the same token, the Human Spirit contains similar ingredients as God, yet it is not God.

The Spirit is like God and it is also a microcosm of the universe. What this means is that every Human Spirit has within itself the potential to be all that there is. People of high attainment such as those who have attained the Spiritual Consciousness Stage are a drop of the universe, in the same way that they are a drop of God. The Spirit being a microcosm of the universe is analogous to the human cell being a microcosm of the human body. That is, the human cell has all the knowledge necessary in its DNA code to create another human body, and by the same token, the Human Spirit has all the knowledge in it to create another universe.

Our Christian Bible states that God created man and woman in His own image and likeness. This is a true statement as long as the interpretation does not mean that God is a physical being and we are like him in that regard. From the description in this book we are created by God and have many of his Attributes but the physical in not one of them.

Chapter 3

Waves

3.1 Introduction

This book advocates that God the Absolute created the universe out of His own mental substance and that that substance is made of vibrating waves. Consequently our universe is a mental universe made of waves. Everything that we could possibly imagine is made of waves to include all matter, energy, and mind. This chapter is an attempt to explain what waves are, how they work in our seen physical universe, and to try giving a basic understanding of how they work in the invisible part of the universe. This chapter provides a basic description of sound waves, electromagnetic waves, and the latest thinking of energy and matter waves from a theoretical scientific prospective. A basic understanding of waves as we know them now, will also help provide an understanding of how the unseen part of the universe works.

The whole universe, at every level of its existence, is made of waves that vibrate at an infinite number of frequency rates and there are no exceptions. For that reason it is important to know something about waves, vibrations, wavelengths, and frequencies. Wave theory is complex, but you don't need to be an expert; however, it is advisable to have, at least, a rudimentary understanding of basic concepts. This presentation is a simplified explanation of wave mechanics to give the average layperson an elementary understanding, without using complex mathematics. Waves make up everything in the universe, this includes the Physical Domain, the Astral Domain, all three of the Spiritual Domains, all the mental planes, all energy, and even the Spirit, Soul and Consciousness.

3.1.1 The Most Fundamental Wave

The *sine wave* is the most fundamental wave in the universe. The basic sine wave is a very simple wave and it is fascinating that the entire universe is formed from these simple sine waves. There are only two basic variables to the sine waves; they are (1) the *rate of vibration (frequency)* and (2) the *amplitude*. These two variables in waves are what create the infinite manifestation of phenomena in the universe. By changing the rate and amplitude of the wave will change the characteristics and properties of the wave, and that, in a nutshell is what gives us the infinite variety of phenomenon in the universe we live in. A pictorial representation of one cycle of the sine wave is shown in the diagram below.

Diagram 2
The Sine Wave

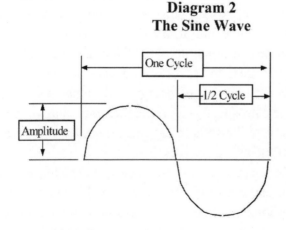

Frequency and amplitude defines the waves. Frequency provides the number of waves per unit of time and amplitude provides the positive and negative height of the wave crest. Frequency is usually denoted as cycles per second, cycles per minute, revolutions per minute, repetitions per hour, etc. The term *cycles per second* has been changed to *Hertz* in honor of Heinrech Rudolph Hertz (1857-1894), a German scientist, who discovered the electromagnetic waves through scientific experimentation in the 1880's.

High school Trigonometry teaches all the basics of the sine wave plus the fundamental principles of wave theory. Trigonometry is the prerequisite course for all future studies in math, science, and engineering. Mathematicians, scientists, and engineers are very familiar with wave theory because it is part of their education and training from the beginning of their studies, and it continues throughout their professional careers. It has only been in relatively recent human history that we have learned that there are such things as waves in every facet of our universe. Ancient people could only see waves in the ocean or vibrating stings in musical instruments. Since the nineteenth century our knowledge of waves has expended tremendously.

3.1.2 Many Types of Waves

Wave theory describes the known physical universe very accurately in mathematical terms. It explains such things as the turning of the wheel, the swing of the pendulum, the orbit of the planets around the sun, the vibration of sound, the transmission of light through space, the transmission of power through power lines, and the functioning of electronics in a computer chip. Some examples of wave vibration in our everyday experiences are as follows:

- Electrical power into the home is measured in terms of 120 volts at 60 Hertz. The 120 volts being the amplitude of the wave, and 60 hertz the frequency of the wave. The wave shape of electrical power into the home is normally a perfect sine wave.

- The rotation of the wheel or the motor rotating at constant speed is almost a perfect sine wave. Rotations of the wheel or motor are commonly denoted in revolutions per minute.
- The clock rates in Personal Computers (PC's) are expressed as mega-hertz and more recently in giga-hertz. Generally digital clocks are square waves, which are complex version of the sine wave.
- Radio waves are measured in kilohertz and television waves are measured in mega-hertz. There is quite a lot of difference in the frequencies of each, and also in the properties and characteristics of each.
- Other examples of motion that are expressed as waves are the rotation of the planets around the sun, and the rotation of the electron around the nucleus of the atom.

The sections that follow in this chapter provide a more in depth discussion on the basics of sound waves, electromagnetic waves, matter waves and mind waves. These explanations are an attempt to show how waves work in the known physical environment, and then apply those same basic principles to try to understand how the unseen universe works. The principles are basically the same in both the seen and the unseen worlds. This is an application of the principle of correspondence, "as below so above, as above so below".

3.2 Sound Waves

The familiar sound waves contain characteristics that are common to all waves of the universe. It is thought that if we learn some of the basic characteristics of sound waves we can begin to understand some of the basics of the waves in the rest of the universe. There are three main characteristics of sound, which we should pay attention to as follows:
1. Frequency – The frequency of sound changes the pitch that we hear. For instance low frequencies create low pitch tones,

which we call base sound and high frequencies produce high pitch sounds. As the frequency changes from low (about 20 cycles per second) to high (about 20,000 cycles per second) the sound changes accordingly. We hear a different pitch of sound for every frequency on the scale from low to high.
2. Amplitude – The amplitude of the wave is what produces the loudness of the sound. Low amplitude creates a soft sound and high amplitude creates a loud sound.
3. Complex waveforms – Sound waves combine in many different ways to produce complex waveforms. Simple sine waves of different frequencies can combine (add or subtract) to form complex waves. Both the frequency component and the amplitude component can combine by adding or subtracting. The complex waveforms produce a different sound from the original simple sine waves that formed the new wave, but the original sine waves retain their unique characteristics.

We cannot see sound waves because they are made of air molecules that compress and decompress in varying degrees of air pressure. Our ear picks up these minute fluctuations in air pressure and the brain/mind interprets them as sound. Voice and music are complex sound waves. Other waves of nature are very different from sound waves but a common trait of all waves is that a change in frequency and amplitude cause a change in the characteristic behavior of the waves. This discussion on sound waves is provided only because sound is familiar to all people and it may helps provide a basic understanding of wave properties.

3.3 Electromagnetic Waves

Electromagnetic waves are all around us all the time. Light is a small portion of the *electromagnetic spectrum*, and of course it is the only part of the electromagnetic spectrum that we can see. Our scientists have learned a lot about electromagnetic waves since the nineteenth century. An elementary study of the electromagnetic

Florencio J. Perez

waves will be sufficient to help in forming a conception of how the waves in the unseen universe work.

Electromagnetic waves were discovered in the second half of the nineteenth century. James Clerk Maxwell predicted their existence through scientific and mathematical calculations in 1864, and Heinrech Hertz discovered these waves experimentally between the years of 1886 and 1888. Since then many scientists and engineers throughout the world have made many new discoveries in rapid succession. Much of our hi-tech and dot-com economy of today is based on the knowledge gained on the behavior of electricity, magnetism, and the electromagnetic wave.

The electromagnetic wave is a waveform composed of an electric field and a magnetic field that are in perpendicular planes to each other. Some of the characteristics of electromagnetic waves are as follows:

- Visible light is made of electromagnetic waves which travels at $c = 186,300$ miles per second. The speed of light is the fastest possible speed in the physical universe.
- The electromagnetic spectrum is an immense range of frequencies.
- Electromagnetic waves of different frequencies and characteristics coexist with each other in the same space.
- Electromagnetic waves reflect, refract, and polarize.
- The properties and characteristics of the electromagnetic spectrum changes as the frequency and amplitude of the waves change.
- The origin of electromagnetic waves is within the atoms of matter. The quantum jump of electrons in the orbits of the atom causes photons to be released or absorbed. Photons are electromagnetic radiation.

3.3.1 Segments of the Electromagnetic Spectrum

The electromagnetic spectrum is grouped into several segments according to their characteristics, properties and material of origin. The list of each of these segments with a brief description and frequency range follows:

- Electric power and Audio - Electrical power in the US is 60 hertz, and audio frequencies ranges between 20 to 20,000 hertz. Audio frequencies, of course, correspond to the range of audible sound frequencies.
- Radio waves - These frequencies are subdivided as follows:
    - Very low frequencies (VLF) range is from 10 to 30 kilo-hertz
    - AM broadcast range 535 to 1,605 kilo-hertz
    - Short wave band range 9 to 21 mega-hertz
    - VHF television bands range from 54 to 216 mega-hertz;
    - FM broadcast band ranges from 88 to 108 mega-hertz. FM is set inside the VHF range.
    - UHF television range is 470 to 890 mega-hertz
- Microwaves range is from 6×10^8 to 1×10^{12} hertz.
- Infrared waves are a narrow bandwidth around 1×10^{13} hertz. These waves are invisible to the naked eye, but our engineers have developed instrumentation that will detect and convert infrared to visible light. These instruments are widely used in the military to see at night. This is so, because live humans are heat sources. Our bodies radiate heat, which can be seen as infrared radiation.
- Visible Light is only a very small portion of the electromagnetic spectrum. It lies between 3.7×10^{14} to 1×10^{15} hertz. The seven colors of the light spectrum (the rainbow) are red, orange, yellow, green, blue, indigo, and violet. The color red corresponds to the lowest frequencies of light, and the color violet corresponds to the highest frequency of light.

- Ultraviolet waves are just beyond the upper limit of visible light at around 1×10^{15} hertz. A common source of ultraviolet light is the sun. Ultraviolet light cannot be seen, but it produces sunburn when the skin is exposed to the sun's rays.
- X-rays extend from about 1×10^{16} to 1×10^{20} hertz. X-rays are high-energy rays that can penetrate through the human body, and are commonly used by medical doctors for medical diagnosis.
- Gamma rays extend from about 1×10^{20} to upwards of 1×10^{24} hertz. Gamma rays are also high-energy rays produced during nuclear reactions and are lethal to the human body.

It should be noted that 1×10^{24} hertz is 1,000,000,000,000,000,000,000,000 cycles per second; that is, a one with 24 zeros. The human mind cannot comprehend that number. The frequency of 1×10^{24} is the highest limit in the perceptible Physical Domain. Our present hi-tech instrumentation cannot observe any frequencies higher than that. We cannot detect any thing higher than gamma ray, but that does not mean that nature stops there.

3.3.2 Broadband

An important characteristic of electromagnetic waves is that they can be modulated with other frequencies. Modulation means that several frequencies can be combined (added or subtracted) and yet the independent characteristics of the original waves are not changed. Microwave frequencies (called carrier frequencies) happen to have the property that they can be transmitted through the atmosphere and picked up by an antenna at some distance away. Radio signals or television programming signals may be modulated into the carrier frequency and transmitted to a receiving unit at home. The definition of modulation by Encarta dictionary is, *"the process of changing the amplitude or frequency of a wave, used in radio broadcasting to superimpose a sound signal on a continuously transmitted carrier wave."*

The significance of broadband is that, because of its relatively high frequencies, a large amount of information can be modulated into it, for transmission. For instance one high frequency carrier frequency (say around 1×10^{11} hertz) can contain within it all the television channels, and carry them to many televisions sets in many homes. The television set at home tunes in to the carrier frequency and then extracts (demodulates) the television information out of the carrier frequency. Light frequencies are much higher rates than microwave frequencies and can therefore carry much more information than microwave. Light frequencies can now be modulated with massive amounts of information, and this information can be transmitted over fiber optic cable. This means that one fiber optic cable can simultaneously carry thousands of telephone conversations, plus many television channels, and Internet data. It is easy to rationalize from this scenario, that the higher the frequency, the higher the amount of intelligent information that can be transmitted or assimilated. This applies also to the storage of data, as in a Compact Disk (CD), or the Digital Video Disk (DVD). The reason why a discussion of high frequencies for broadband is important is that frequencies at the Spiritual Domains are even higher than those of the Physical Domain. Consequently information in the Spiritual Domains is more efficiently transmitted, stored, and assimilated. This principle is significant to our overall discussion in later chapters.

3.3.3 Coexistence of Frequencies

Another important characteristic of electromagnetic waves is their ability to coexist in the same space at the same time, regardless of their frequency or intensity. For instance, you can have several flashlights, each with a different colored filter (different frequencies) in a closed room and you can shine all of them at a wall. You can form separate spots on the wall with the different colored flashlights or you can mix the colors in any combination. They all coexist in the same space without interfering with the original properties.

This characteristic applies whether it is light waves, radio waves, microwave, or gamma rays. It also applies to the ultra-high frequencies of the unseen domains that we will be discussing later on.

3.4 Matter and Energy Waves

It is a well-known scientific principle that matter and energy are interchangeable. This is evident from Einstein's equation, $E = Mc^2$, where E = energy, M = mass, and c^2 = the speed of light squared. This matter-energy equivalence rule states that matter is stored energy. Also evident from the equation is that a small amount of matter contains a huge amount of energy. An example of this is the nuclear bomb, which releases gigantic amounts of energy when a small amount of uranium is detonated. It is also a well-known scientific fact that energy is made of waves and it follows that matter is also made of waves. Tremendous gains were made in the twentieth century towards understanding the nature of matter and energy. These theories all rely on mathematics to probe into the inter secrets of the universe and at the core of these investigation is wave theory. In the last century science discovered the paradox of the wave/particles duality of both energy and matter. That is, both energy and matter has the properties of behaving as either waves or particles. A brief and simplified description of some of the most prominent theories is provided.

3.4.1 Atomic Structure

Niels Bohr, a Danish physicist, proposed the model of the atom in 1913. Bohr's model is use today but only in a general sense, Schrödinger's model is becoming more acceptable as the understanding of the atom continues to develop. Bohr's model proposes that the atom is composed of a nucleus at the center of the atom and the electrons orbit the nucleus. The nucleus is composed of protons and neutrons. Electrons have a negative electrical charge and stand-alone. Protons have a positive charge, and the neutrons are electrically neutral. In the early 1920's many scientists thought they had the solution to the fundamental makeup of matter. This assumption, however, did not last long. The Quantum Theory was developed in the mid 1920s and it led to the discovery of a whole array of new subatomic particles within the proton and the neutron. It was learned that all these subatomic particles also have the same particles/wave duality syndrome, depending on the condition but the dilemma is always best resolved in favor of waves.

We normally perceive matter as being solid, liquid, or gas with properties such as hard or soft, hot or cold, heavy or light, course or fine, bright or dark, sticky or non-sticky, and so on. It is hard for us to think of matter as waves especially when matter is made up mostly of empty space. It turns out that the micro-universe is very different from the macro-universe that we are most familiar with.

To explain this we need to look into the structure of the atom just a little further. Referring to Bohr's model of the atom, the size of the atom is extremely small and there is a substantial gap between the nucleus and the electron orbiting the nucleus. In a hydrogen atom, which is the smallest and simplest atom in the Physical Domain, the electron orbits the nucleus at a distance of around 10,000 times the radius of the nucleus. If we were to blow up the scale of the hydrogen atom to a size we can see, we make the nucleus the size of a basketball, and the electron would be the size of a pinhead. The distance of the pinhead revolving around the basketball would be at a radius of 15 miles. In this example there would be nothing but empty space between the basketball and the pinhead. From this illustration it becomes intuitively evident that atoms (and matter in general) are made mostly of empty space. To add to this mystery, it should be noted that neither the electron or the nucleus are solid objects. They are made of waves that vibrate at different frequencies and there is nothing physical about them. There is nothing tangible about them; they are only mathematical things.

3.4.2 Quantum Mechanics

Quantum mechanics is also known as quantum theory or quantum physics. The name quantum has to do with the fact that an electron can jump from one orbital level of the atom to another orbital level and this is referred to as a quantum jump. A quantum jump of the electron releases or absorbs photons, which are a unit of energy. The photons are electromagnetic radiation (light), which are made of waves but can also behave as particles. Light bends when it passes close to a large celestial body with large gravitational pull such as the sun because of the small amount of mass contained in the photon.

Quantum Mechanics is a set of mathematical equations that were developed in the mid 1920's to further explain the inner working of matter. In 1926 an Austrian physicist, Erwin Shrödinger, developed the set of mathematical equations known as "wave mechanics". At about the same time, the German mathematician Werner Heisenberg independently developed a set of equations that were very different from Schrödinger's, yet they explained the same phenomenon very well. Heisenberg's equations were known as "matrix mechanics" and their work became the basis for Quantum Mechanics. Both of these methods, seemed different at first glance but upon further investigation by independent scientists it was discovered that the two methods were essentially equal. They both explain matter in terms of waves that vibrate at many different frequencies. Since then there have been much advancement on the theory of Quantum Mechanics.

The discovery of Quantum Mechanics has made it possible to test and explain the results of many other theories of matter. It underlies virtually every modern invention in electronics, telecommunications, computers and nuclear energy. Quantum Mechanics forms the basis for the new physics and has made possible the transistors, lasers, radar, microchips, television, and many other hi-tech devices that we take for granted today.

One of the many byproducts of the development of Quantum Mechanics was that the equations were able to predict that the proton and neutron were made of many smaller particles. These subdivisions are called *subatomic particles* or *elementary particles*, and many of them have been verified experimentally in *cyclotrons* or *particle accelerators*. From the standpoint of this book, the important thing to remember is that the quantum equations prove that all subatomic particles and indeed all matter and energy in the universe are made of waves.

3.4.3 Excerpt from Schrödinger's Lecture

Erwin Shrödinger won the Nobel Prize in 1926 for his work that provided the mathematical equations for Quantum Mechanics. In 1952 he delivered a lecture in Geneva, Switzerland that discussed the apparent paradox of the duality of matter, which behaves as both a wave and a particle. A condensed version appeared in *Scientific America* the following year. This excerpt is from *Microsoft Encarta 2001 Encyclopedia* article "What is Matter?"

What is Matter?

"Fifty years ago science seemed on the road to a clear-cut answer to the ancient question which is the title of this article. It looked as if matter would be reduced at last to its ultimate building blocks—to certain submicroscopic but nevertheless tangible and measurable particles. But it proved to be less simple than that. Today a physicist no longer can distinguish significantly between matter and something else. We no longer contrast matter with forces or fields of force as different entities; we know now that these concepts must be merged. It is true that we speak of "empty" space (*i.e.*, space free of matter), but space is never really empty, because even in the remotest voids of the universe there is always starlight—and *that* is matter. Besides, space is filled with gravitational fields, and according to Einstein, gravity and inertia cannot very well be separated."

"Thus the subject of this article is in fact the total picture of space-time reality as envisaged by physics. We have to admit that our conception of material reality today is more wavering and uncertain than it has been for a long time. We know a great many interesting details, and learn new ones every week. But to construct a clear, easily comprehensible picture on which all physicists would agree—that is simply impossible. Physics stands at a grave crisis of ideas. In the face of this crisis, many maintain that no objective picture of reality is possible. However, the optimists among us (of whom I consider myself one) look upon this view as a philosophical extravagance born of despair. We hope that the present fluctuations of thinking are only indications of an upheaval of old beliefs, which in the end will lead to something better than the mess of formulas which today surrounds our subject."

~~~~~~~~~~~~~~~~~~~~~~~~~~~~~~

Science thrives in explaining all phenomena in an objective manner; it is apparent from this excerpt that objective proof for all their theories may be going by the wayside. Consider of Schrödinger's statement, "We have to admit that our conception of material reality today is more wavering and uncertain than it has been for a long time." and also the statement, "Physics stands at a grave crisis of ideas. In the face of this crisis, many maintain that no objective picture of reality is possible." What this is saying is that there is now a blur between the objective (physical) and the subjective (nonphysical). From the standpoint of classical physics, all theories must stand the scrutiny of empirical testing. As modern theories extend beyond the possibility of physical tests, they become subjective, and this is the realm of metaphysics. This book of course goes beyond the physical and explains phenomena of reality in the realm of the metaphysical.

3.4.4 Subatomic Particles

As a result of the development of Quantum Mechanics there have been over one hundred fifty subatomic particles identified. Some of these subatomic particles are known as muons, neutrinos, fermions, hebrons, masons, pions, kaons, photons, gravitons, gluons and etc. The most fundamental particles are identified as *(1) electrons, (2) leptons, (3) quarks, and (4) bosons.* These are weird names for strange little things that make up our Physical Domain. The electrons are independent and revolve around the nucleus; all the other particles reside in the nucleus. That is, the nucleus is made of protons and neutrons and they in turn are made of subatomic particles. Leptons and quarks have many variations with unfamiliar names. Bosons are transmitters of fundamental forces that cause subatomic particles to interact with one another, and to keep them organized. There are four fundamental forces of matter as follows:

1. The strong nuclear force - responsible for forces between quarks
2. The electric charge - responsible for electromagnetic radiation
3. The weak nuclear force - responsible for some radioactive processes.
4. Gravity - Gravity is relatively unimportant for subatomic particles because it is much weaker than the other three forces, but it is very powerful in the macro-universe.

There are many subatomic particles that do not occur naturally in atoms. Some of these particles have been predicted to exist through mathematical deductions, while others have actually been produced and observed in particle accelerators. One important set of these subatomic particles is *antimatter*.

3.4.5 Antimatter

Antimatter does not exist naturally in the physical universe but it can be produced in particle accelerator with high-energy collisions. Scientists have discovered that for every subatomic particle there is a corresponding anti-subatomic particle. The anti-particles have the same mass and energy as ordinary matter but they have opposite electrical charges. For example, the negatively charged ordinary electron has a corresponding anti-electron called a positron, which has a positive charge. The positively charged ordinary proton has a corresponding anti-proton, which is negatively charged.

These suggest that along side our ordinary visible universe there is an invisible anti-matter universe. This means that there are invisible anti-galaxies, anti-suns, anti-worlds, anti-plants, anti-animals and anti-humans. That is, for every ordinary human physical body there is a corresponding human body composed of antimatter. Both bodies coexist side by side or superimposed upon each other at all times. The anti-matter particles that scientists observe in laboratory experiments are of extremely short duration (in the order of micro-seconds and/or neno-seconds). Hypothetically, anti-particles are not stable in the Physical Domain because they are not of the Physical Domain; they are of the Astral Domain where they are indeed stable. The stable version of antimatter is what scientists are now calling *dark matter* and this book calls astral matter because it is the matter that pertains to the Astral Domain.

3.4.6 Dark Matter and Dark Energy

In recent years scientists have deduced that there is some other kind of unseen matter and energy in the universe. These deductions come about from scientific calculations that reveal that the total gravity estimates of the universe do not correspond with the total amount of detectable physical matter estimates in the universe. The logical presumption is that there is something else out there that we cannot see, and because it cannot be seen it must be dark, hence the name *"dark matter"* and *"dark energy"*. In metaphysics this has been known all along. In this book dark matter is called astral matter, and dark energy is called astral energy, which forms the material that makes up the Astral Domain. The Physical Domain and the Astral Domain coexists in the same space in a superimpose manner. They coexist in the same space because they are made of different frequencies that coexist without interfering with each other. This is similar to the light from several flashlights that coexist in the same room.

3.4.7 Super-string Theory

Super-string theory is a recent development in the new physics that is an extension of the quantum theory. Super-strings are tiny little vibrating strings (metaphorically speaking) that are the basic building material for all matter and energy. In very simplistic terms, this theory states that everything in the universe (all subatomic particles, all forces, and perhaps even time and space) consist of fantastically small strings that vibrate. Of course these strings are not really physical strings; the term string is used only as a metaphorical illustration. The mathematics of strings and super-strings are very complex; it involves multi-dimensions of strings that spin and form extremely small loops that occupy infinitely small space. Super strings are also an attempt to unify all four fundamental forces into one single force. There is no physical counterpart to use as reference since there is no physical equivalence.

A crude illustration can be used, however, which equates the *super-string* to the strings of a guitar or violin. In a guitar for instance, when a string is plucked, it vibrates causing a sound, and by changing the length of the string there is a corresponding change in the pitch of the sound. The *super-strings* also vibrate, and each super-string has a characteristic, which corresponds to the frequency of its vibration. The characteristic of the *super-string* will change if the rate of vibration changes. This suggests that subatomic particles may change characteristics by changing vibrating frequencies. It also suggests that the universe is made of only one kind of string, and the difference in vibration is what gives each string the different characteristics. These strings vibrating at different frequencies are what make up our physical universe, and upon further reasoning we can deduce that the universe is not as solid as we perceive it. Even though the super-string theory has been successful in explaining many properties of matter, there is still much theoretical work to be done.

The significance of the super-string theory is that everything physical (matter and energy) is made of vibrating substance, which corresponds well with the teachings of metaphysics.

3.5 The Mental Universe

It has been shown above that matter is made of concentrated energy per Einstein's equation. It is the contention of metaphysical teachings that energy is made of a concentration of mind. Of course, we do not have an equation yet to explain the relationship between mind and energy. The implication here is that the universe is made of mind substance, which goes along with the earlier statement that the universe is created from Gods mental substance. In essence we live in a mental universe.

A hypothesis for the most basic material for the construction of mind follows.

3.5.1 The Wavelet Hypothesis

At the present time the Superstring Theory is the strongest contender to ultimately explain the true nature of matter and energy. I propose the "Wavelet Hypothesis" to describe the phenomenon of mind, which lies outside the realm of physical matter and energy. This hypothesis is actually an extension of the Superstring Theory. Wavelets are envisioned to be exceedingly small and vibrating at extremely high rates. It is like very short piece of a ray of light. A ray of light in this case is like a very long one-dimensional string that vibrates at the frequency of light depending on the color of light. A metaphor is used here to try to present a mental picture of what a *wavelet* might look like. In actuality a wavelet cannot be seen and there is no chance of anyone ever seeing one. This is just an educated guess, which seems to meet the requirements to explain the basic building blocks of mind. The wavelet is real, as real as real can be, but it is not real in an objective sense; it is only real in a subjective sense, which turns out to be more real than the objective sense. It is similar to saying God is real as real can be but He is not a physical object. He is only subjective and we can only imagine Him in our mind's eye. It is only through intellectual effort, logical deductions, faith and a number of other things that we believe in God. It is not because we can see Him physically. We build our own subjective evidence of God and base our philosophy on that personal evidence. Mother Nature is of God and we can think of Mother Nature as an expression of God in the physical, if we so desire. We are relative beings at our particular stage of development and objective things

Evolution of Consciousness

are real to us. As we develop further we will eventually realize that what we call subjective now belongs to the absolute realm of God, which in affect is more real than the objective that we are familiar with.

3.5.2 Metaphor of a Wavelet

Suppose that you could take a ray of light and cut it into many small pieces with a special pair of scissors. Imagine that you had a ray of red light, a ray of blue light, a ray of green light, a ray of violet light and so on. Then you take one ray at a time and cut it up into many little bitty segments with the special scissors, and as you cut you put these little segments into a special jar that can contain the segments. These little segments are always cut at the full node of the wave, never at a fraction of a wave, and each segment may have between ten and a hundred complete waves in it. The little segments continue to vibrate in the special jar at the same rate as the original ray of light. These little vibrating segments are now called "wavelets". Suppose you continue to cut all the rays (the red, blue, green, and violet rays) with the special scissors and you put them all in the special jar. When you finish cutting you end up with a jar full of wavelets that are of slightly different lengths and vibrating at many different rates. These wavelets in the jar are vibrating at the rate of different colors of light, but now let us suppose that these wavelets increase their rate of vibration by a factor of a trillion, and now they start to reach the rate of wavelets that are the building blocks for mind.

3.5.3 Properties of Wavelets

Some generic properties of the hypothetical wavelets are as follows:

- The wavelets vibrate perpetually, and they vibrate at much higher rates than light. There is an infinite spectrum of frequencies for wavelets.

- Each wavelet coexists in the same volume of space (figuratively since there is no space and time at this realm) with other wavelets, with each maintaining its own unique characteristic.
- Wavelets have a type of mass, inertia, and charge which is akin to physical matter but wavelets are not physical. Wavelets cannot be detected or measured by physical means.
- Wavelets are equivalent to subatomic particles in physical matter, but again wavelets must not be confused with physical matter.
- Wavelets of similar categories unite to form complex Mental Elements, which are similar and akin to atoms and molecules of the physical type. Independent wavelets have their own characteristics but when they combine with other wavelets to form Mental Elements the characteristic changes. In other words, the characteristics of the wavelet and the characteristic of the element are different, but the wavelet imbedded in the element has not changed. An example of this is white light, which can be separated into its constituent colors of the rainbow. When separated they appear as distinct colors but when combined they appear as white light, yet the independent colors have not changed. Another example of this is water H_2O that is made of two atoms of hydrogen to one atom of oxygen. Independent oxygen and independent hydrogen have their own characteristics, which are very different than that of water.
- Wavelets coexist with physical material with no interference.

3.5.4 Building Material of Mind

Wavelets are the basic building material of mind. Wavelets are equivalent to subatomic particles and they tend to coalesce to form complex *mental elements*. Mental elements are extremely small units of mind that have intelligence and themselves congeal in an infinite number of ways to form more complex *mental molecules*. Mental molecules are somewhat equivalent to chemical molecules in the Physical Domain. Mental molecules form the basic building blocks for all the varieties of mind in the universe, which exists everywhere in the universe. It exists in every domain and in every plane of every domain. Wavelets in the lower domains are of relatively lower rates of vibration and become progressively higher rates for each successively higher domain. For instance, mental wavelets in the Spiritual Domain are made of the highest frequencies of the universe.

All minds at all domain planes are modular. The special jar in the example provided above is intended to represent a module of mind that is formed of wavelets. Wavelets form mental elements, mental elements form mental molecules, and mental molecules form mental modules. Mental Modules form mental systems and subsystems. A tabulated list from the smallest to the largest is as follows:

- Mental wavelets
- Mental elements
- Mental molecules
- Mental modules
- Mental subsystems
- Mental systems

There is mind to control every function of the universe from the most primitive form of inanimate material to the most advanced form of intelligent life. Universal Mind is the mastermind of the universe and under its supervision are all the billions and billions of minds that perform all the billions and billions of functions of the universe. Most of the mind in the universe is of the instinctive type, which performs specific repetitive functions in an almost automatic mode with very little supervision from the mastermind. Some of the most prevalent instinctive minds listed in hierarchical from lower to higher are:

- Mineral Instinctive Operating Mind (MIOM)
- Terrestrial Instinctive Operating Mind (TIOM)
- Galactic Instinctive Operating Mind (GIOM)
- Plant Instinctive Operating Mind (PIOM)
- Animal Instinctive Operating Mind (AIOM)
- Human Instinctive Operating Mind (HIOM)

These are very prevalent instinctive minds and are treated in more detail in later chapters; however there are many more types of mind in the universe. Other types of mind that are discussed extensively are the Human Intellectual Mind, Human Personality, and Human Intuitive Mind.

3.5.5 Building Material of Consciousness

Consciousness is akin to mind but it is not mind. Mind is a tool to the Consciousness and is not permanent. The basic building material for Consciousness is also a form of wavelet that is of a special form, which is similar but different to that of the mind wavelet. It has the capacity to accumulate huge quantities of mental wavelets as necessary throughout its evolutionary and growing process.

Consciousness is a relative entity but gradually evolves to absoluteness. It has the quality of being the Real Self of the human being and it is permanent, everlasting, and eternal. The Human Consciousness is the same Consciousness and therefore the same person throughout the evolutionary process of that person.

3.5.6 Building Material of Spirit and Soul

Spirit and Soul are made of "God's Divine Spark", but this is only because of lack of better words. God is unknowable because He is infinite and absolute. We can only assume that Spirit and Soul are made of a form of wavelet but this is strictly speculation on our part. Spirit and Soul are absolute entities, which is a quality of God.

Chapter 4

The Law of Cause and Effect

Universal Laws govern the universe and they apply equally everywhere in the universe, that is, they apply in the Physical Domain, as well as in Astral Domain, as well as in the three Spiritual Domains. Universal Laws are immutable, inflexible, unbreakable, and eternal; they apply to all aspects of the universe including matter, energy, mind and Consciousness. The universe and the Universal Laws are creations of God, the Absolute, but the universe is a relative universe whereas the Universal Laws are absolute. That is, the universe is constantly changing and evolving whereas the Universal Laws that govern it are constant and unchangeable. Universal Laws have been the same since the beginning of time and will continue to be the same forever more. The laws do not change but they do vary in degree throughout the hierarchical plan of the universe. This book expounds on three main Universal Laws. These laws are:

- Law of Cause and Effect
- Law of Relativity
- Law of Evolution

Of these the Law of Cause and Effect is considered to be the master law and will be discussed in this chapter. All Universal Laws are intrinsically interrelated; we separate them only for the sake of explanation. The universe being relative is continually changing and evolving, and by the same token, the universe evolves because it is relative. Relativity is explained in Chapter 5 and evolution is explained in Chapter 6.

Our scientific community knows and understands these laws well, but they only know them as they apply to the realm of the Physical Domain. This book contends that they apply not only to the Physical Domain but also to the Astral Domain and to the Spiritual Domains as well. These laws apply to every nook and cranny of the universe, nothing is ever left out, and there is no escape from the laws. They apply every moment of every day of our lives, and in every moment and region of the universe's existence. Localized laws are sub-laws of the main Universal Laws and apply only to specific domains. For instance most of the laws of physics, which we are familiar with, only apply to the Physical Domain. The Astral Domain has many of its own laws, which are similar to the laws of physics, but different. A law that does not apply to all domains equally is not a Universal Law. An example of this kind of law is the law of gravity, which applies to the Physical and Astral Domains but does not apply to the Spiritual Domains since the Spiritual Domains do not have matter and inertia. Many of the laws of physics such as those that apply to thermodynamics, electricity, dynamics (Newton's Laws of Motion), fluid mechanics, optics and others are not Universal Laws because they apply only to the Physical Domain. The Spiritual Domains also have laws of their own that apply specific to them, but these laws are of the mental realm and not of the physical realm.

Humanity was totally unaware of the laws of Cause and Effect, Relativity, and Evolution until only recently in our history. Isaac Newton discovered the laws of motion in the late 1600's. Charles Darwin discovered the Law of Evolution in the mid to late 1800's, and Albert Einstein discovered relativity in the early 1900's. These laws as explained by Newton, Darwin, and Einstein apply to the Physical Domain only, but this book will endeavor to explain their affect in all domains.

The universe and Universal Laws were designed specifically to propel and guide us through every phase and every stage of our evolution; they were created for the singular purpose of providing the means for Consciousnesses like us to evolve. These laws propel us forward by constantly pushing and driving us onward, and they guide us by not allowing us to deviate much from the proscribed path of development. It is apparent from these statements that we are very important to God, and we should learn to appreciate it. If these laws are so important why are we so unknowing of them, why are we so oblivious of their affect on our lives? We are the way we are because of these laws, yet we are so ignorant of them. Hopefully we can clarify this dilemma in the forthcoming chapters of this book.

4.1 Law of Cause and Effect

The Law of Cause and Effect governs every action or event that has ever taken place in the universe. A simplistic way of stating the law is that "for every action there is an equal and opposite reaction". An example is that if you push on something with some amount of force, it pushes back at you with the same amount of force. This example implies that the law is active in the Physical Domain only, but that is not so; it is always active in the Spiritual, mental, physical, and astral aspects of the universe as well. The human being, through its human mind is the biggest creator of cause and effect. Every thing that we call *man-made* is first initiated in the human Consciousness and then created in the human mind. The human mind then creates thought, which creates action, and action is the sphere of influence of the Law of Cause and Effect. The Consciousness is a part of the Spiritual makeup of the human. The Real Self of every person is composed of the combination of Spirit, Soul, and Consciousness.

4.1.1 Chain of Command for Action

At our level of development even the simplest and most mundane activity, involves the Spiritual, the mental, and the physical parts of the person. The Consciousness is the originator of action and the chain of events follows a proscribed sequence. The chain of command for human action is from top to bottom as depicted below:

Chain of Command

- Consciousness – Initiates the action by providing general instructions to the Intellectual Mind.
- Intellectual Mind – Provides reasoning, analysis, and commands to the HIOM.
- HIOM – Provides detailed mental commands to the brain.
- Brain – Provides detailed electrochemical commands to the nervous system.
- Nervous System – Provides distribution of electrochemical signals to the muscles.
- Muscles – Provides the physical action.

The completion of the action does not mean that the process of the action is completed. Mother Nature has a means of recording the action and then designing the proper reaction to the action. The reaction is the effect of the initial cause. A more complex manifestation of the same principle is when, through your actions, you do harm to another person, whether knowingly or unknowingly, and the reaction of Mother Nature (the effect) is that you will be hurt in return. Mother Nature keeps track of all causes, and each cause requires a corresponding counterbalance with a proper effect. For every cause there has to be an effect; there are no exceptions.

4.1.2 The Human Mind and Cause and Effect

All man made things, big and small, are first initiated by the Consciousness, then created in the mind, and then manifested in the physical. This includes creating such things as a novel, a painting, a super highway, a building, a musical composition, a huge battleship, or the Space Shuttle. The creation could be by one person as in the case of a novel, or it could be the creation of thousands of persons as in the case of the Space Shuttle. In all cases, the thinking process is the cause and the physical manifestation is the effect. Human thought can be good or bad. Mother Nature is impartial when it comes to the subject of good or bad. If the cause is good the effect is good, and if the cause is bad the effect is bad. Good thoughts produce good effects, whereas bad thoughts produce bad effects. Such things as schools, hospitals, businesses, churches, and thousands of other things are produced by good thoughts. Bad thoughts, on the other hand produce such things as theft, murder, terrorism, war atrocities and thousands of other evil things. All thoughts and subsequent actions are accountable to the Law of Cause and Effect. Good thoughts that produce good things are rewarded, whereas bad thoughts that produce bad things will generate pain and suffering. This is the way of Mother Nature; it is designed for our benefit that is the way we learn and move forward in our constant pursuit of personal progress. This is the way Mother Nature, through Her Law of Cause and Effect, propels and guides us forward in our development.

Most bad things produced by humans are produced because of ignorance. For instance we first started polluting the earth because of ignorance. We pollute our earth by many different means such as auto emissions, herbicides, pesticides, trash, atomic waist, and many others. The latest form of pollution is space junk in the space immediately surrounding our earth. At first we were unaware of what we were doing, and did not realize the consequence. We used to think that the earth was so big, so infinitely big, that it could absorb and withstand anything that we had to dish out. At first these were honest errors on our part, since we were ignorant and unaware of the consequences. Initially we started with small quantities of pollutants and did not notice the effects. It was not until later years that we started to notice that we were killing the fish in the lakes and bays, and animals in the forests, and causing health problems for ourselves. Out of ignorance we can do bad things, but the Law of Cause and Effect is always present to remind us. Ignorance is not an excuse but it is a part of human nature. We are here to do things, to make mistakes, and most important of all, to learn from those mistakes. In simplistic terms, that is a part of the process of Evolution of Consciousness.

4.1.3 Imaginary Story of a Little Girl and a Hot Grill

This is a simple little story that can provide significant lessons related to cause and effect. Suppose that a little girl (two years old or so) is playing in the backyard while her father is grilling hamburgers in the charcoal grill. While the grilling is in progress the telephone rings in the house and the father tells the little girl "I am going to answer the phone, you stay away from the grill because it is hot." The little girl being very inquisitive and not knowing the meaning of the word hot goes to the grill and touches it and burns her hand. The result is pain, of course.

At least four lessons can be learned from this simple story as follows:
1. The little girl touched the hot grill because of inquisitive ignorance.
2. The effect was pain, but the pain is not due to punishment, it is due to the law in action.
3. The result is an experience gained and a lesson learned.
4. This lesson will contribute to the development of the little girl, and it takes millions upon millions of such lessons to move forward.

Some people may interpret touching of the hot grill as a *sin*, especially since the little girl disobeyed her father, and the pain suffered may be interpreted as the punishment for committing the *sin*. These interpretations are completely erroneous. In the first place the term *sin* has the connotation that something evil has been committed. This little girl did not commit anything evil; it is perfectly normal to be ignorant of some things. The pain suffered is not punishment; it is merely the normal effect of the cause. What counts in the end is that the lesson be learned.

4.1.4 Cause and Effect of Inanimate Things

Cause and effect occurs with no human intervention as well. For example, the extremely complex phenomenon of weather is all cause and effect. Weather is caused by the sun's uneven heating to the earth's surface which causes hot air to raise causing areas of atmospheric high and low pressure. Air then moves from high to low pressure areas causing wind. The dynamics of weather are very complex and meteorological science is still trying figure how it all works. All natural phenomenon such as earthquakes, volcanoes, mountain building and erosion follow the rules of the Law of Cause with no human thought involved. Actually all earthly activities of this sort take place in the higher Spiritual Domains and manifest in the physical aspects of the universe. Activities in the Spiritual Domains are made of Spiritual and mental characteristic which indeed affect the Physical Domain.

4.1.5 The Laws of Motion

Sir Isaac Newton provided us with an indication of the workings of the Law of Cause and Effect from a scientific standpoint. His discoveries, known as the laws of motion, apply to a narrow portion of the Law of Cause and Effect. Newton built on the previous works of Galileo, Descartes and others, and over the years his discoveries have had a profound affect on the scientific and engineering view of the world. These laws are credited with much of the huge advances in modern engineering and technology. He published his works on the "laws of motion" in 1686 in his book titled, *Prencipia Mathematica*. His three laws of motion describe action and reaction which are functions of the main law of Cause and Effect. They also form the most fundamental natural laws of classical mechanics. Briefly, the three laws of motion are stated as follows.

Law of Motion Number 1, - Inertia – Inertia states that an object in a state of rest or in uniform linear motion tends to remain in such a state until acted upon by an external force.

Law of Motion Number 2, - Force, Mass and Acceleration – Stated as, $F = ma$, where "F" represents force; "m" represents mass; and "a" represents acceleration. Force represents the cause and acceleration of the mass represents the effect.

Law of Motion Number 3, - Action and Reaction – Every action produces an equal and opposite reaction.

These three laws of motion define only a small subset of the total Law of Cause and Effect. The Law of Cause and Effect applies everywhere in Nature, but we only have a few basic mathematical equations to describe it. Science has developed equations to define electricity, fluid mechanics, thermodynamics, optics and more but they only apply to the Physical Domain.

4.1.6 Cause and Effect from Beginning of Time to End of Time

Another important fact of the Law of Cause and Effect is that it has been in operation from the beginning of earth time and will continue forever. The following scenario can illustrate the principle of law in action over the passage of time.

Each one of us is the offspring of two parents. We are also the offspring of four grandparents who conceived our parents. We are also the offspring of eight great grandparents who conceived our grandparents. We are also the offspring of sixteen great, great grandparents who conceived our great grandparents. We are also the offspring of thirty-two great, great, great grandparents who conceived our great, great grandparents. This scenario can go on and on into the past. By the same token, this same scenario also goes forward on and on into the future with our forthcoming offspring.

4.2 The Game of Life

4.2.1 Actions are the Result of Thought

The real effect of action is actually the effect of thought since action follows thought. We are what we are today, simply because we have done, or left undone, certain things in our past lives. We have had certain desires, and have acted upon them, and the result is manifested today. This does not mean that we are being punished because we have done certain things in the past. This means that if our thoughts were good our actions were good and we were rewarded, and by the same token if we had bad thoughts our action were bad and we suffered pain. It means that if we put our fingers into the fire we will have to nurse our burns. The things we did in the past were not necessarily bad things. We merely may have become unduly attached to certain things, and our attachments and desires brought certain effects, which may have been unpleasant and painful. Yet, all things we do are good because they all teach us vital lessons. Once we realize the nature of our problem we learn from it and the hurt will soon fade away. Cause and Effect is not a law of punishment or Divine revenge, it is an equalizer and a deterrent factor to learn from.

4.2.2 Work for the Sake of Work

Humans are instinctively motivated to action, to doing things, and to work. Action is instinctively in our nature. We should stress a routine of hard work in our lives since work is in our nature anyway. By action we mean doing things that are practical and fruitful such as holding a regular job for wages, going to school for an education, obtaining professional training, participating in sports, hobbies and anything that causes learning through experience. As the Consciousness unfolds and awakens, through experience after experience, it gradually obtains an intelligent and true nature of itself and of its quest.

An inactive or idle person does not make mistakes and consequently does not make progress. Idleness is not recommended because it is a waste of time, and in severe cases it could mean a squandered lifetime. Stress the need for hard work; because work is to be enjoyed, but do not let work become stressful. We stress the need to work because work brings new experiences, knowledge, wisdom, and personal development.

Our responsibility is always to act and participate in the great game of life. We are to do the best that we know how and do it to the best of our ability but at the same time remembering that we should not become attached to the result of our action. The idea is to work for the sake of work and not for fame or fortune. We are to do our work gladly, cheerfully, willingly, and with gusto, but at the same time we have to realize that the material gains as a result of hard work are relative things and consequently of no real value. The value of work is in the learning and not in the material wealth that it brings. This statement seems like a bad contradiction to the way most of us think therefore it requires an explanation. In reality the explanation is very complex and beyond our level to understand, nevertheless an effort is made to offer a crude explanation.

One of the first things that we have to realize is that each one of us is a unit in the whole machinery or plan of life. We each have our place in this great plan and we must each do our job to the best of our ability. No matter how important a position we may have, or how responsible our job, we are all but units in the Divine Plan and we must be willing to be used in accordance with that plan. By the same token, no matter how lowly or unimportant we may feel, we are still a unit in the same plan and have a purpose and a job to perform within that plan. Nothing is unimportant, the most advanced as well as the less advanced people are equally subject to the laws of that plan. We must all play our parts as well as we know how, not only because we are working out our own development and evolution, but because we are being used by a Divine mind as a pawn in the great game of life. This does not mean that we are mere android, far from that; it means that our interests are bound up with the rest of humanity, and we touch all mankind sooner or later. We must learn to be perfectly willing to be used by this Divine mind, and thus we will find that this willingness will prevent friction and pain.

The universe is interrelated, interconnected, and interactive and our actions affect others while the actions of others affect us. The result of actions may be small and affect a few people or it could be profound and affect a nation. For example, if you wash dishes, mop the floor, mow the lawn, and wash the car you affect the members of your family. However, the actions of Martin Luther King in the 1960s were profound since they affected the course of a nation for years to come. The interconnectivity of the world is especially apparent in modern days with the advent of globalization where we have global trade, global banking, global labor shifts and much more. It is now practical for a car to be designed in Germany by German engineers, and assembled in America by American labor using Japanese manufactured parts.

Our lives are not just for our own personal development, but also for the development of the entire human race. A certain job may seem useless or insignificant to our personal development, but it may be important in the grand scheme of things (the Divine Plan) and we should do our job willingly. Every thing we do has a meaning and sooner or later that meaning will become clear. This is so whether we know it or not; it is like a move in a chess game that may seem meaningless at first but is shown to be an important move later in the game. Such is the case in the actions we take; the actions we take will be seen as important moves later in the grand scheme of life's plan. We must learn to be moved by the Spirit without complaint, knowing that all will be well in the end, and that the move is needed to effect certain combinations or changes in the noble game of life, which all of humanity is playing. It is difficult for most people who do not understand the inner workings of the game to allow to be guided by the unseen Spirit, and they normally impart resistance to the plan. This resistance however causes friction, which in turn causes grief and suffering to the person refusing to accept. The advanced person who understands the secret and is willing to be guided by the Spirit will have a much easier time. Generally the people who learn to trust the Spirit will escape the grief and reap positive benefits. It is not uncommon for successful people to rely on the Spirit and be moved by it to the point where they do little on their own. Often some of these successful people are thought by others to be smart, gifted, and superior, but unbeknown to them is the fact that these successful people only provide the action through which the Spirit works.

4.2.3 Work, Success, Wealth and Happiness

It is always our desire to seek happiness. The problem is that we tend to seek happiness outside of ourselves rather than within ourselves. If a person is asked "What does happiness really mean to you?" The response is usually that they want to be successful at what they do with the ultimate goal of becoming prosperous so they can own material things such as a nice home, automobile, and a substantial bank account. These are commendable goals but these people often find that happiness is always lurking beyond their reach. This is apparent in some people who suddenly come into quick wealth, and they soon find that this wealth does not bring the happiness they dreamed. There is nothing wrong with acquiring wealth, provided that you learn to remain unattached because material things are relative and subject to change and impermanence. The secret is to work for the sake of work without becoming attached to the material fruits of your labor. You must remain aware that you can not take material things with, but you will take all lessons learned as a result of gaining material wealth. It does not mean that a person should not enjoy the material gains; it only means that the person should not grow too attached to those material things. If the principle of hard work for the sake of work is followed faithfully, the material wealth will come automatically and will even multiply; it is to be used and enjoyed without forming strong attachments.

The difference lies in the fact that the attached person believes that his or her happiness depends on certain things or persons, while the freed person believes that happiness comes from within rather than from outside. The freed person is able to convert things to pleasure, which would otherwise produce dissatisfaction and disappointment. As long as one is tied to, and dependent upon any particular person or thing for happiness then he or she is a slave to that particular person or thing. But when a person is free of entangling influences, then that person is master of his or her own destiny and happiness. This does not mean that we should not love others, on the contrary we should love others generously but the love should be unselfish, unconditional and unattached.

If we seek out some expected source of happiness and eventually find it, we will most likely find that it carries the sting of pain. But if we stop looking at the thing as a source of happiness and instead look at it as an accomplishment of life then the poison is neutralized and the sting is blunted. Greed is the result of the false assumption that "more stuff" (at any cost) is happiness, but material stuff cannot be happiness because it is relative and relative things, on the long run, are "here today and gone tomorrow". If you look for fame and fortune as the thing that will bring happiness, you will most likely find it and grow attached to it, but you will also find that the price of attachment is anguish without joy. To the one who is freed and who works for the love of work without becoming attached, fame and/or fortune will come as an incident and the anguish will not be experienced. There are many things to which we devote our entire lives and usually we bring in more pain than happiness. We tend to look for material thing for happiness rather than to look within for happiness. The moment that we look for happiness on an outside things or persons we open the door to heartache and unhappiness. This is so because outside things or persons cannot satisfy the internal longing of the Consciousness. The Consciousness cannot be dependent on outside sources for its happiness because outside sources are relative entities, which are not permanent. It has to depend on itself (the Real Self) because that is where the Divine part of itself is. Our Divine part is our Soul and Spirit which are absolute entities; they are true reality and they are permanent.

4.3 Sub-Laws of the Law of Cause and Effect

The Law of Cause and Effect is a master Universal Law, but it can be broken down into sub-laws that are also universal in nature because they apply to all domains, aspect, and realms of the universe. An excellent book on this subject is written by Bruce McArthur and is titled, *"Your Life - Why It Is the Way It Is and What You Can Do About It – Understanding the Universal Laws"*. Mr. McArthur did an exceptional job of explaining the Law of Cause and Effect and its sub-laws and also in explaining how they affect our lives.

Only a few of the sub-laws are mentioned here, and the "Law of Attractions" is referred to throughout the book because of its predominance. Like everything else in nature, all sub-laws are complex and extend from the relatively simple to understand to the incomprehensible. Often, a thorough study and deep thought is required to appreciate the beauty and profound significance of the workings of these sub-laws in our lives.

4.3.1 "Law of Attractions" stated as "Like Attracts Like"

This law deals with the tendency for attraction of like things, which includes attraction between individuals, thoughts, situations, and conditions. The Law of Attractions is a sub-law of the master Law of Cause and Effect and it is universal because it is prevalent in every domain and facet of the universe; it is prevalent in the physical, mental, and Spiritual aspects. A few examples of the law in action are as follows:

- People with like thoughts and beliefs attract each other.
- When you search, you will find. This is so because the law attracts you towards what it is you seek, especially if it fulfills your needs. The law makes it such that you are drawn towards such things as workshops, books, schools, people, or discussions that you are in tune with. Jesus eloquently said this on the Sermon on the Mount, (Mathew 7-7) *"Ask and it will be given to you; seek and you will find; knock and the door will be opened to you. For everyone who asks receives; he who seeks finds; and to him who knocks, the door will be opened"*.
- A common adage that portrays the law is, "birds of a feather flock together", which means that "like attracts like". This applies to animals of the same species, to humans, to human thoughts, to human customs, to common age groups, to religious groups, to philosophical groups, to entertainment, to emotions and much more.

- During rebirth into an earth life, the Soul-Consciousness chooses its parents. It is attracted to parents that are of like-mind, like-pursuits, and equivalent level of development; it is also attracted to parents that will fulfill its desires and aspirations for growth.

What you think is what you do and also what you attract towards yourself. Examples of positive emotional attractions that can be engendered if you act in a certain way are as follows:

- If you are gentle you will attract gentleness toward yourself.
- If you are kind you will attract kindness toward yourself.
- If you are friendly you will attract friendship toward yourself.
- If you are truthful you will attract truth toward yourself.
- If you are good you will attract decency and goodness toward yourself.
- If you are joyful you will attract joy toward yourself.
- If you are honest you will attract honesty and integrity toward yourself.
- If you are just you will attract justice toward yourself.
- If you are forgiving you will attract forgiveness toward yourself.
- If you are generous you will attract abundance toward yourself.

This list can go on and on, but it must be understood that the law is impartial. Just as you can use the law to attract positive attributes toward yourself by thinking positive, you can also attract negative attributes if your thoughts are negative. For instance, some negative attitudes that can be attracted toward you can be as follows:

- If you are cruel you will attract cruelty towards yourself.
- If you are unforgiving you will attract intolerance or vindictiveness toward yourself.
- If you are deceitful you will attract dishonesty or untrustworthiness toward yourself.
- If you are unfair you will attract injustice toward yourself.
- If you are sad you will attract more sadness, gloominess, or grief toward yourself.

Again, you can change your life by changing your attitude. Negative attitudes can be converted to positive attitudes by thinking, acting, and talking in the manner required, and it takes self-discipline to do that. Self-discipline is the buzzword to keep in mind. Positive attitudes can be enhanced to even greater heights by continuing to think in terms of even higher positive levels of that attitude. All attitudes and emotions can always go to even higher degrees and higher levels of the present condition. The highest manifestation of all emotions and attitudes is a transformation to true love, more appropriately known as "Divine Love".

4.3.2 "Law of Like Conditions" stated as "Like Produces Like"

This law in its simplistic form can be seen in the fact that plants and animals reproduce after their own kind. For instance dogs produce dogs, roses produce roses, pine trees produce pine trees, and humans produce humans, etc.

The same sub-law also applies to all human aspects of living such as thoughts, actions, situations, and emotions. For instance:

- Emotions produces like emotions
- Attitudes produces like attitudes
- Thoughts produces like thoughts
- Motives produces like motives

- Acts produces like acts
- Peace produces peace
- Friendship produces friends
- Joy produces joyfulness
- Love produces love

On the negative side of the equation:

- Hate produces hate
- Bigotry produces bigotry

This is a clear indication of the fact that you create that which is in your mind, the law is unbiased and you can create either positive or negative situations according to the nature of your thoughts. Another way to look at it is that you are the master of your own destiny; no one is responsible for your destiny but yourself. Universal laws are designed in such a way that they give you the responsibility for creating your own fate. If you do not like your present life, you have the responsibility and the tools to change your life to a more desirable life style. There are two requirements needed to change your life: (1) first you have to become aware of your problem, and (2) you have to change your life style by changing your thoughts through force of will, desire, resolve, and self-discipline.

4.3.3 "Law of Increases" also stated "As you sow, so shall you reap".

Nature has a way of multiplying things. If you plant a few watermelon seeds in the right kind of soil you can grow many watermelons with thousands more seeds than what you started with. These law of increases applies to plants as well as to animals; for instance, bacteria, ants, locusts, salmon, sardines, rats, rabbits and many other organisms can multiply in huge quantities if conditions are right for them.

If you plant a good thought in fertile ground it would grow and multiply in the minds of other likeminded people. Bruce McArthur writes, *"Another important point is that we sow these seeds not only by spoken words of acts toward others, but also by thoughts. Our thoughts are basic to our creative power; we use them to direct our energy. We can consider the thought as the seed that carries all the potential of that thought, just as a wheat seed carries all the potential of the plant it will produce."*

4.4 Veil of Ignorance

In our primitive levels of development we were completely ignorant of our world. We were created knowing nothing of our environment. It has taken us thousands of years in our evolutionary process to reach the point where we are now. It seems strange that we have to start out ignorant when someone already knows everything there is to know about our universe. The universe is a knowable entity, and as a matter of fact there are many highly developed Consciousnesses who know all that there is to know in the universe.

However, even though we were created knowing nothing, we were also created with the necessary set of tools to allow us to learn and develop; therefore we were also given the capability to conquer ignorance. These tools are the human physical body with its powerful brain, and the *personal mind* that we each possess. Each human *personal mind* consists of (1) the Human Instinctive Operating Mind (HIOM), (2) the Personality, (3) the Intellectual Mind, and (3) the Intuitive Mind. The combination of these tools gives us that "Innate Intelligence", which is instrumental and indispensable in our development. The other important tools include the "earth school" that we live in, and the Universal Laws that propel us and guide us through the evolutionary process.

The following is an illustration intended to help us understand an important aspect of the Law of Cause and Effect, and the influence it has on our evolution. This illustration is in the form of a narrative to aid in explaining something that is difficult to understand. Stories can often clarify things.

4.4.1 A Narrative on the "Veils of Ignorance"

In this narrative, suppose there is a light bulb that is about four inches in diameter, which produces a bright white light, and inclosing the light bulb, are millions of opaque veils that cover and smother out the light. A veil is a lightweight piece of cloth that women sometimes wear to cover their head. It usually has gaps in the material that allow some light to pass through. In this illustration there are enough veils to form a spherical ball that is ten feet in diameter.

Now, suppose that the light bulb represents the Consciousness and each veil represents an ignorance that smothers the light of the Consciousness. Therefore these veils are called "veils of ignorance", and they keep the Consciousness from shining its light to the world. We do not commit errors because we are inherently bad people. The reason we commit errors is because we are ignorant of a particular situation. One way to remove ignorance is to do things, commit errors, experience the result of the errors (per the Law of Cause and Effect), and then correct the errors. The end outcome of correcting the errors is that we learn a lesson, we gain knowledge, we expand our wisdom, and we remove the ignorance. In other words, we replace ignorance with knowledge. All errors that we commit, correct, and learn from, are represented by "veils of ignorance" that are removed and replaced by "veils of knowledge". A veil of knowledge is completely transparent in this story. It takes thousands upon thousands and even millions of errors committed and lessons learned to gradually remove those millions of veils of ignorance. A veil of ignorance is like a layer of limitation that requires removal; removal represents a step forward in our evolutionary journey.

It takes a long time to laboriously remove each of those veils of ignorance; it actually takes tens of thousands of years and many, many lifetimes to accomplish this. A veil is not removed unless the lesson is fully and totally learned, and there is absolutely no way to fool Mother Nature. If the lesson is not learned the first time, it has to be repeated until it is truly learned. Mother Nature is infinity wise, and don't think for a moment that you can deceive Her. Some people may tend to think that no one can ever know what they are thinking; their thoughts are their own, and as long as they do not talk about them, no body will ever know their secrets. Well, that is not so; all thoughts are made of waves that are imbedded in energy and form thought clouds that float in the astral atmosphere. (There will be more detail on this subject in subsequent chapters.) Those thought clouds surround the person thinking them, and they cannot be hidden from Mother Nature. As long as the person has negative thoughts, that person cannot make progress up the path. Each negative thought is a veil of ignorance that needs to be replaced with a veil of knowledge. It is a common misconception, especially by cynics, that if something is in the mind, it can't be real. You may hear words such as, "Oh, it's all in your mind" meaning it is not real. In a truer sense, if something is in the mind it is real. All physical manifestations first occur in the mind.

Once a lesson is learned well, it is never forgotten. The opaque part of the ten-foot ball in our illustration gradually gets smaller and smaller, and eventually the light of the Consciousness begin to penetrate through the confines of the remaining veils of ignorance. The veils of knowledge, being transparent, allow the light to get through and light up the world. The principle of this illustration is an example of how the Law of Cause and Effect guides us in our ever-present quest for personal development.

There are many ways to erase ignorance, and replace it with knowledge; Mother Nature does not care how we learn as long as we learn the lessons well. This illustration describes the brute force method of learning by doing and committing errors. Another method that is a much easier way to learn is through formal education and/or training. Education and training can contribute tremendously to the learning process and alleviate much of the consternation of pain through the effects of errors. A person can also learn by watching and following others who have already learned the lessons.

From the standpoint of this book we are not *sinners* with the connotation of being bad, we are just ignorant and we make mistakes, but always with the goal of personal growth. The Law of Cause and Effect is always present, never failing, and always guiding us in our pursuit. Evolution of Consciousness is the process of replacing ignorance with knowledge. Note that no one can learn a lesson for someone else; each person has to experience and learn everything on their own; there are no short cuts to the process.

When we first start off as primitive humans, we are completely ignorant of our surrounding environment, but gradually we advance through the ranks of development. Each one of us has been the same individual throughout all these eons of time. We gradually advanced from the primitive state to the advanced state that we are in now. We have never changed identity through time; we have only changed our level of awareness. And yet, we are still only scratching the surface of the advances yet to come. The hard part of our evolution is now in our past for most of us; the future is easier and holds many wondrous and glorious things for us, but it is still long and arduous. We still have a long way to go. It is like we are now reaching the peak of a range of mountains that we have been striving to climb for a long time, and as we reach the top we sigh with relief only to be disappointed again. As we look over the top of the mountain range we just finished climbing we see that there is another mountain range that is even higher than the one we just completed. At that time, we realize that we now have to climb the new mountain range as well, and that the journey is much longer than we had ever imagined; we also become aware that the road ahead is still full of strife. It may seem depressing at first but we have to recognize that there are more wanders, glories, and joys to be experienced as we go up the mountain. Evolution is hard work at every stage of our development but it is also sprinkled with a lot of wonderful satisfactions. The final satisfaction is the reunion with God Himself at the end of the long journey.

4.5 The Cycle of Human Life

As mentioned previously, the Law of Cause and Effect is a universal law because it has an influence in all parts and in all aspects of the universe. The law is prevalent everywhere and there is no way to escape it. This includes all domains as well as the Spiritual, mental and energy aspects. All causes are originated in the Spiritual and the effects can be as diverse as the infinity of the universe. Evolution of Consciousness is the accumulation of experiences, which becomes a database of knowledge that translates into wisdom. The Law of Cause and Effect is the main law that guides and propels Evolution of Consciousness. A principle that is also essential and works in conjunction with the Law of Cause and Effect is the *"Principle of Reincarnation"*. The two are inseparable in our present level of development. Evolution of Consciousness would not be possible without the principle of reincarnation. The whole purpose of this wonderfully created universe is to provide a means for Consciousnesses like ours to develop, unfold, and evolve.

4.5.1 Reincarnation

The principle of reincarnation is not obvious to the human five physical senses; therefore many skeptics argue that reincarnation is a make-believe figment of the imagination. This book treats reincarnation as a true and necessary principle of the universe. It is interwoven and interrelated with other topics of this book. It is interrelated with the subject of relativity and as well as evolution, so a brief general introduction is presented here as an educative edification. More detailed explanations will be provided throughout the book as it applies to the subject matter. Reincarnation applies only to people of the Human Consciousness Stage and the Soul Consciousness Stage because they live physical lives on earth. It does not apply to the Cosmic Consciousness Stage and the Spiritual Consciousness Stage since they do not have a need to live earth lives anymore.

The ability to live many lives is what gives us the capability to gain many experiences, and develop and evolve forward. Evolution is the process of accumulating thousands upon thousands or even millions upon millions of little experiences and formatting them as a foundation of wisdom. These millions of experiences cannot be gained in one lifetime. It takes many, many lifetimes to accumulate enough knowledge to make us as smart as we are now. It is impossible to be born knowing everything that we know presently without the benefit of the experiences gained over of many lifetimes. If this was our first and only lifetime on earth we would be about as smart as the caveman of old. Life is a cycle that we repeat over and over again, and with each life cycle we progress forward. The cycle includes living in the physical earth environment and then transitioning to live in the non-physical Spiritual environment. In simplified form the steps of the cycle are as follows:

- A person is born into earth life as a baby. This means that the Soul-Consciousness enters the infant's body at birth.
- A person lives through infancy.
- A person lives through early childhood.
- A person lives through puberty.
- A person lives through adolescence and teenage years.
- A person lives through early adulthood.
- A person lives through middle adulthood.
- A person lives through senior adulthood.
- A person lives through old age.
- A person dies a natural death. This book uses the term death because it is the commonly used expression in our society, but perhaps a better term to use is transition. Death tends to denote finality of a person, and that is not so.
- A person transitions to a period of rest for a relatively short duration.
- A person moves to a plane within Spiritual Domain 1 (SD1) where the person has the same level of awareness as in the previous earth life. There is no change in the person's mental composition from the experiences in earth life. The person lives in this environment for many years, in which the person continues to develop in every respect similar to earth years.
- The Soul and the Consciousness together determine that it is time for another earth life and start the process of rebirth.
- The Soul-Consciousness enters a new baby's body at the time of birth and starts a new physical life. This is *reincarnation*. Notice that the physical body is temporary but

the Soul-Consciousness is forever and ever. It is always the same Soul-Consciousness that acquires a new body for each new life on earth. Nothing can ever destroy the Soul or the Consciousness.

Why are we so unaware of the reincarnation process? One reason is that the reincarnation phenomena operate outside the range of our five physical senses; another reason is that Mother Nature has designed a mechanism where we forget things of the past. We remember recent events much better than long past events. We forget the specific events, but the experiences gained by those events are cumulative, and are stored in a subconscious database that is part of the Consciousness. In our present lifetime, for instance, we all forget the events of our babyhood. Yet the experiences of that timeframe are very important in our lives because they remain indelibly in our subconscious, and they play an important role in forming our present personalities. It seems that as we get older we forget more and more of the particulars of our past. For instance, I even forgot the name of my date to my senior high school prom. Just think, if we cannot remember many of the events in our present life, how can we be expected to remember specific events of past lives that occurred, perhaps, hundreds of years previous to our present lifetime.

There are various important reasons why Mother Nature built in that veil of forgetfulness in our human way of being. For one reason, Mother Nature is not interested in the specifics of the events; the important thing is the cumulative affect of the learned essentials; of the knowledge acquired as a result of the events. Specific events are relative and do not mean much but the experiences obtained are real since they serve a purpose and are retained forever in the database of the Consciousness. It would not be well for most people to remember details of previous lives because they may not be able to bear some of those recollections. Think of how some people in the present life foolishly grieve for long extended duration over a loved one that passed away. Think of those who remain angry with a past boss who may have fired them, or, those who may remain bitter over a "dear John letter" or a divorce. There are some who may be mentally wounded as a result of some horrific experience such as the act of terrorism, or war, or tragic accident. If we could remember the specifics of the events of many past lives we would be incapable of dealing with the present life. Mother Nature is wise to impose the veil of forgetfulness on us.

4.5.2 Leading Events towards Rebirth

The decision to be reborn for a new lifetime on earth is a combination of many factors. Some of the factors involved for instance are: 1) The completion of a life in the Spiritual Domain; 2) the need for additional earthly experiences for the sake of growth; 3) a natural urge exists to return to the material world again; 4) links due to past causes and effects; and a multitude of other reasons which are too numerous to know. Both the Consciousness and the Soul have an important role to play in selecting the parents for the next lifetime, and of course, the "Law of Attractions" also plays its role through the whole process. The Consciousness tends to look for parents and the environment that it is familiar with. It must select parents that have similar personality traits and similar level of development, otherwise strife would be likely through the baby's development. It perhaps looks for parents that it knew in a previous life and that have similar likes and dislikes, or similar aspirations and goals. It looks to fulfill its longings at that point in its evolutionary process.

The Soul has a similar but different prospective. The Soul looks for parents that will provide the best possible opportunity for further development of the Consciousness. It may seem that there are conflicts between the Soul and the Consciousness when making the choice for new parents, but the conflict is more apparent than real because an equitable compromise is always reached. The final decision is always made amicably.

The time that a person spends living in Spiritual Domain 1 (SD1) between lifetimes varies from person to person. Normally the less advanced people stay for relatively short duration while the more advanced people stay for much longer durations. People that live in our present day society may stay in SD1 for from 50 to 150 earth years. Often it may be thousands of years between incarnations, but remember that SD1 does not have time, as we know it on earth. So when a Consciousness is reborn into a new earth life it has to face a different world from the one it knew in the previous lifetime. The new world has changed in many, many ways simply due to the progression of history on earth. For instance, there are changes in technology, in institutions, in morals, in dress, in finance and in almost everything you can think of. The Consciousness usually changes nationality, language, race, religion, sex, profession, and goals, from one lifetime to the next. These changes are always designed to maximize the potential for growth, yet enough similarities must remain (like attracts like) to maintain continuity. This book can only present an oversimplification for a general understanding of this very complex subject of reincarnation.

4.5.3 Reincarnation Answers Nagging Questions

Knowing the principle of reincarnation can often help explain many nagging social questions or riddles in our daily lives. An explanation to some of our present day social questions follows:

- **Homosexuality** – Homosexuality often but not always, occurs when a person changes sex from one lifetime to the next. For instance, if a person is a female in the previous lifetime and changes to being a male in the present lifetime, then this person will likely be attracted to people of the same sex in this lifetime. This is so because as a female in the previous lifetime she liked men, which of course was normal, but with the change in sex to male in this lifetime it continued to like men because strong personality traits or instincts tend to continue. The problem is not in the physical makeup of the brain, but in the mental makeup of the person. Changing sex from one lifetime to the next is a normal phenomenon of nature; it has existed throughout history and shall continue. Like everything else in nature, the decision as to the sex of

the newborn baby is very complex, but it involves the balancing of the requirements of cause and effect as well as the needs of the Soul and the Consciousness.
- **Time of Birth** – The Consciousness enters the baby at the time of birth. Usually it enters at the time of the first breath but it could also be delayed a few hours. Occasionally the Consciousness may take longer than that to enter the baby, and may cause other complications in the baby's life.
- **Waking up** – The Consciousness remains in a semi-slumber condition throughout the childhood years. It does not fully awaken until about the time of puberty or early adulthood. Normally the Consciousness wakes up completely by age 18 of the person.
- **Balancing Cause and Effect** – Cause and Effect remains influential in the person from lifetime to lifetime. Causes in a previous lifetime will most likely produce effects in the present lifetime. Some people may grumble about having bad luck or jump-for-joy with gook luck, but the reality is that many things happen to a person as a result of natures urge to balance out the cause and effect of the person. Nothing happens by chance, but the process is so complex that there is no way we can understand or predict the outcome.
- **Heredity** – Heredity between the parent and child only pertains to physical features due to the effects of DNA. Personality and mental traits are not inherited between parent and child. The only reason personality characteristics seem to match between parent and child is because the Consciousness did a good job of picking its parents during its initial search. Remember that the Consciousness was searching for parents with similar characteristics to satisfy the Law of Attractions as well as looking out for its own comfort level.
- **Database** – All the events and experiences of the present life are recorded in a subconscious database, which is a

part of the Consciousness. All events of past lives are also recorded in the same database, but they are located in deeper stratums of the subconscious and therefore more difficult to access. Even though we do not directly recall details of the experiences recorded in the database we are in fact a product of the content of the database. The experiences recorded in the database remain there forever but tend to fade away to the background as they are replaced by new and more advanced experiences.

4.6 Spiritual Cause and Effect

The Spirit and Soul are Divine parts of us and they do not change because they are absolute entities, but the Consciousness is constantly in the process of changing. It is continually active, doing things, and making mistakes, which leads to the process of learning new lessons and gaining knowledge, and wisdom. This process of learning by making mistakes goes on from lifetime to lifetime, or incarnation to incarnation. Sooner or later the Consciousness learns that some activities are hurtful and discovers the folly of certain actions. It learns that certain ways of living are bad and will eventually learn to avoid the wrong actions.

Most of us have noticed in our present live that some wicked things are tempting to us while other wicked things are not as tempting. That is, we have outgrown the lure of some wicked things, but are still tempted or enticed by other wicked things. The reason for this is that those wicked things that tempt us are things that we have not personally experienced in the past and we have a curious desire to see for ourselves. Doing these wicked things, of course, bring us nothing but hurt and pain. The pain is what sets us to think about our behavior and enlightens us to the cause of the pain; thereby eventually we learn the stupidity of the violation and learn to refrain from continuing the infractions. A lesson, when truly learned, is never forgotten, even from lifetime to lifetime. So, when you find that certain temptations do not bother you, you can be sure that you have already learned this temptation. On the other hand, when you see a friend that cannot see the error of certain actions, when it is clear to you, then you should realize that you have already gone through that stage, while your friend has not. Your friend is just now going through the pain that you have already experienced in the past, and which you have already outgrown. You do not realize that in a future life, your friend having learned the lessons by experience will be free of the hurt and pain, just as you are free now.

It is hard to realize that we are who we are as a result of our experiences. You may sometimes wish you could eliminate some painful experience from your life, or some disgraceful episode, or some humiliating occurrence. But, you have to remember that if it were possible to get rid of these things, you would be forced to part with the experience and knowledge that has made you who you are now. If you were to be able to eradicate those experiences, it would be like going back to the way you used to be in the past, and you would probably be inclined to live those horrible experiences any way. The bottom line is, do not try to eliminate earned experiences even if they were painful; instead try to learn from them, and try to never repeat them again.

You may say, "Of what use are the experiences gained in former lives, if we do not remember them? They are lost to us." But they are not lost to you, they are built into your mental structure, they are yours forever, and nothing can ever take them away from you. Your character now is based on the accumulation of all your past experiences, which include those of all past lives as well as those of the present life. You remember some of the important present life experiences, but forget details of experiences from past lives. The resulting wisdom of the accumulated experiences stays with you forever, whether they are recent experiences that you remember well, or old experiences that have faded from your memory. All experiences are cumulative and are woven into your mental fabric forever. It is these experiences, which give tendencies that make you like or dislike certain thing. They influence you to choose some directions over some other directions. These influences may cause you to instinctively recognize certain things as being wrong or unwise, and consequently cause you to turn away from them. The totality of past experiences is what gives you your present inclinations for likes and dislikes, interpretation of good and evil, and feelings about such things as love, beauty, kindness, etc. Nothing is lost in life, all the experiences of the past contribute to your well being in the present, and all your troubles and pains of the present will bear fruit in the future. We do not always learn the lesson on the first go round, and we may have to repeat the lesson several times until we learn it, however, nothing is ever lost and every time we repeat a lesson it becomes clearer and easier. Details of past actions are lost just as material wealth is lost with passage of time or death, this is so because they are relative things and therefore not permanent, but the wisdom gained due to the actions taken will never be lost.

The doctrine of Spiritual Cause and Effect is based upon the great truth that under the law, each person is master of his or her own destiny. Each person is its own judge and jury consequently each person renders its own rewards as well as its own punishment. Each thought, word, or action has its effect on the life of the person, and this is not as a form of reward or punishment, but as a result of the Law of Cause and Effect. Good thoughts, words or actions produce rewards whereas bad thoughts, words or actions produce punishment as a result of the law and not because there is a policeman watching over you.

The Soul knows what is really best for the Consciousness. When the Soul senses that the Consciousness is moving in the wrong direction and running wild the Soul tries to swing it around to a better course, or to bring it to a sudden stop if necessary. This is not done as a form of punishment to the Consciousness but as a guidance, protection and kindness. The Soul is not an outside power, the Soul and Consciousness together form the Real Self of the person but of course the Soul is the Divine part of the real person. The Soul is in touch with the great overruling Divine Intelligence, which is God the Absolute. The Soul is an absolute entity and does not change, it does not develop and evolve, whereas the Consciousness is a relative entity that changes and evolves all the time. The Soul does not stop the Consciousness on its tracks because it wants to be mean or has a feeling of righteous indignation. The Soul stops the

Consciousness from hurting itself because it is the right thing to do. The Soul is like a thoughtful parent who is looking out for the child and has the best interest of the child in mind. A wise parent will guide the child along through life, allowing the child to make mistakes for the sake of learning from those mistakes, but if the child is ever in any danger the parent quickly steps in to protect the child. It is like a parent who removes a knife or a gun from a child to keep the child from hurting itself. The Soul will put a stop to the wrong doings by the Consciousness even if the Consciousness tries to rebel or throws a temper tantrum just like a child normally would. But in the end, the Consciousness will listen, obey, and abide by the assertions of the Soul, just like a child abides by the wishes of the disciplining parent. Sometimes the Consciousness does not want to abide by the wishes of the Soul, and that can only lead to much pain and suffering. But sooner or later the Consciousness comes to its senses and understanding its foolishness, repents, and gets back on track.

We have to learn that the Soul's purpose in guiding us is for our ultimate good. The Soul is always trying to balance our experiences. We have to learn that no matter how adverse or cruel the conditions seem they are exactly what we need to balance our lives. The conditions imposed by the Soul are for our greater advantage. We may need strengthening in certain areas of our lives to round us up, and we may be given a nudge by the Soul in that direction. Or we may be tending too strongly in one direction and we are urged in another direction in order to keep a balance in our lives. This is done in a subconscious realm but if you listen you can become aware. This process of listening is really not difficult. For instance, when you are ready to do something you should first stop to listen to your feelings, and if you feel comfortable doing it then it is OK, but if you feel uneasy or unsure, that means your Soul is guiding you and you should refrain from doing it.

4.6.1 Relationships

Spiritual Cause and Effect, in part, also governs our relationships with other people throughout our many lives. In our past lives we have formed attachments with other people, either through love or hate, or through kindness or cruelty. And because of the Law of Attractions these relationships tend to continue in our present lives and tend toward mutual development, advancement, and adjustment. The law dictates that if we hurt someone in our past life we will probably receive hurt by that same person in this life. By the same token if we offered kindness to someone in a past life we are likely to receive kindness by that person in this life. Cause and effect is not a law of revenge but of balance.

An illustration that can help explain the above scenario is as follows. Suppose that in a past life a boy deliberately won the love of a girl for selfish reasons, and then after gratifying his urges deliberately dumped her and made fun of her, causing her a lot of pain and suffering. It is likely that in the next life these two people will come together once again because of the Law of Attractions but this time it will be pay-back-time. The likely scenario this time will be that the boy will fall madly in love with the girl, and the girl will show no affection in return. The boy will suffer pain as a consequence of loving in vain. This is the law in operation, but it is not for the sake of taking revenge. The underlying purpose in this case is to teach the lesson that the sacredness of human affection should not be maliciously tampered with.

Those people whom we loved or were friends with in past lives are likely to be a part of our present life because of the workings of the Law of Attractions. When we experience a sudden like or dislike of a total stranger, it is probable that we were acquaintances in a past life and those emotions carry in our subconscious. We are constantly bound up with the lives of others for pain or happiness, and the law of Spiritual Cause and Effect must run its course. There is a method to escape the wrath of having to pay for all of the infractions of the past or present lives, and that is to gain knowledge of the truth and develop and evolve our Consciousness to higher levels. Every time you advance up the ladder of evolution you change your wave vibration to higher frequency, and the higher frequencies are dominant over the lower frequencies. Therefore you can outgrow and become dominant over older causes of lower frequencies by

Evolution of Consciousness

advancing yourself to higher frequencies. In other words, Mother Nature will forgo some of your infractions and you do not have to pay for all your wrongdoings of the past, provided you outgrow the level of the infraction.

Mother Nature is not trying to punish you for your infractions; She only wants you to learn your lessons. Of course, telling you to learn your lessons is easier said than done. Generally, what you have to do to advance is to try to figure out what your shortcomings are, and then through deep thought and meditation, you can then reverse those shortcomings by repenting. That is, changing your attitude and your way of thinking towards the positive will help you grow in awareness. You have to positively change your ways internally; you cannot just pay lip service to the situation. Mother Nature does not make mistakes and you cannot fool Her in any way. Unbeknownst to you, when you change your ways, your vibrating frequencies also changes and this frequencies are like a beckon to Mother Nature; She knows exactly where you stand at all times by monitoring that beacon. When you change to the positive attitudes your rate of vibration also changes to higher frequencies. You can learn lessons the hard way by committing infractions and paying for them with pain, or you can learn the easy way by changing your attitude through prayer, meditation and education. The Law of Cause and Effect is extremely complex and for the most part beyond our comprehension, but never the less we must learn to accept it. If you cannot accept the concept, try to think about it, experiment, and eventually you will see that it makes sense.

Chapter 5

Relativity

5.1 Introduction to Classical Relativity

The principle of relativity was obscure and basically unknown until the 20th century, when in 1905 Albert Einstein first published the *"Special Theory of Relativity"* and then in 1916 he published the *"General Theory of Relativity"*. These theories, when first disclosed, were not understood or accepted by the scientific community and much less by the general public. Gradually over a period of several years, the theories started to explain physical phenomenon that could not be explained previously. Experiments were conducted, over a period of time that started to clarify and prove that the theories were correct. Now these theories are well accepted and over the years have had profound significance in the thinking of the scientific community.

Relativity remains generally unknown to the general public even though some that are aware of it tend pay little or no attention to its existence. The relativity that was explained by Einstein is outside the range of comprehension for most people because it is outside the range of our five senses, which are the tools we have to perceive the outside world. In order to truly understand Einstein's concepts a person requires an enormous amount of training in physics and mathematics. Yet relativity is always present in our world, and is constantly active in our lives. We live with it every day of our lives and don't know it. The reader is not required to be fully versed in relativity, but an understanding of some of the fundamental principles is needed. This explanation is not intended as a course in relativity. It is only intended to give the reader the basic ideas, enough to show the reader how pervasive relativity really is. It is everywhere, all the time, and we cannot escape it. It plays an extremely important part in our lives and in the general make up of who we are, and why we are the way we are.

Einstein's relativity is referred to as a theory; this book upgrades that concept and calls it a Universal Law. Relativity is a Universal Law because it is real and it is prevalent in every domain and facet of the universe. We live in a relative universe. Does this mean that the universe is an illusion? Well, yes and no. Our Spirit and Soul are absolute entities, but our Consciousness and the rest of the universe are relative. It is proper to say that we live in a *"relative reality"*. From the standpoint of an "absolute reality" our "relative reality" is an illusion, but for us our "relative reality" is real and that's as it should be.

Prior to the publication of the theories of relativity by Einstein, the scientific community accepted the concepts of time, space, motion, matter, and gravity as described by Isaac Newton, which are now known as the *classical laws of physics*. Newton's basic principles (Laws of Motion) are briefly discussed in the previous chapter on Cause and Effect. In the Newtonian mechanics, space and time are considered absolute. Absolute space is homogeneous and is the same everywhere; there is no difference between any two points in space. Absolute time is also the same everywhere in the universe regardless of the frame of reference. Newton's principles work well as long as the objects are of moderate size and are moving at moderate speeds. These principles are still being used very effectively in many branches of engineering and in some branches of astronomy. Einstein's theory of relativity changed Newton's concept in that time and space are no longer absolute, but relative. According to Einstein, space has a curvature around massive bodies with huge gravitational fields, and time slows down as speed approaches the speed of light.

5.1.1 Einstein's Special Theories of Relativity

Einstein's *"Special Theory of Relativity"* describes the relationship of matter, time, and space, showing that mass and energy are equivalent, and that mass, length, and time change with velocity. The theory deals with motion that is in a straight line and at a constant velocity. Observers in different frames of reference and moving at different uniform velocities with respect to each other, would measure space and time differently from each other, yet the laws of mechanics are the same in both frames of reference. According to Einstein, everything in the universes is in motion with respect to something else. All motion is relative and there is no absolute point of reference. For instance, the train is in motion relative to the train station, which appears to be at rest, but the station is not actually at rest since it is moving as the earth spins on its axis and also orbits around the sun.

5.1.2 Einstein's General Theory of Relativity

Einstein's *"General Theory of Relativity"* deals with gravity and how gravity correlates to acceleration. He discovered that the forces produced by gravity and those produced by acceleration are equivalent and indistinguishable. He also deduced that a massive star or planet would bend light as the light passed close to the planet or sun. By the same token a high rate of acceleration would bend light. By association, if gravity bends light, then gravity must also bend space, consequently he deduced that outer space is curved.

The term "Space-Time Continuum" was coined to describe a location in space in terms of space and time. That is, there are three dimensions of space (length, width, and height) and one dimension of time, which now represents a four-dimensional coordinate. Space-time continuum can be used as a measure of four-dimensional space. By Einstein's account, space is not a linear function and by no means is it absolute. Space is curved or distorted by the presence of massive bodies with massive gravitational force. He formulated that massive bodies cause a curvature of space and developed a special geometry of the time-space continuum to define the characteristics of curved space. He tossed out the concept of gravity and replaced it with the concept of curved space.

A "black hole" is the result of a massive star that used up all its nuclear fuel and decayed as a result and then collapsed into a relatively small space with huge gravitational forces. Black holes are thought to exist in all galaxies. The massive gravitational vortexes of the black hole sucks in all matter and light that comes into close proximity and also produce huge curvatures and irregularities in space. Only physicists with extensive training in physics and mathematics can clearly understand these concepts. As for the rest of us, we have to accept the concepts on faith, but to many these concepts are too bizarre to dwell on. We do know that most of these theories have been tested through scientific experimentation and proven to be correct. General relativity is of great scientific importance, and is now used to accurately figure out the motion of cosmological bodies for the purpose of space travel.

5.1.3 Facts of Relativity

The following is a list of factual deductions, which have been made as a result of knowledge gained about relativity through mathematical and experimental means.

- There is no absolute motion, only relative motion. Events that may appear simultaneous to an observer in one system may not appear simultaneous to an observer in another system.
- The velocity of light is a constant in nature and not dependent on the motion of the source of the light.
- The velocity of light is the fastest and limiting velocity in nature. No energy can be transmitted at a velocity greater than that of light (186,000 miles per hour).

- Objects appear to contract in the direction of motion. At the speed of light the object would contract to zero length, thus space contract to zero space, and therefore there is no such thing as space at the speed of light. This is a difficult concept for our mind to understand and accept.
- The mass of a body in motion is a function of the energy content of the body and varies with the velocity. The higher the velocity, the higher the energy of the body. At the speed of light the mass of the body would be infinite and therefore impossible.
- Matter and energy are equivalent by Einstein's famous equation: Energy is equal to the mass of the body times the speed of light squared, ($E = MC^2$). The speed of light being a very large number means that a small amount of mass contains a huge amount of energy. Another way to express this is that mass (matter) is like a storage battery for energy.
- Time is relative. The rate of a moving clock slows down as the velocity of the clock approaches the speed of light. At the speed of light there is no such thing as time. For example, we think that five minutes is a set amount of time that does not change no-matter-what, but in actuality five minutes in a space ship traveling at close to the speed of light, is not the same as five minutes for a person standing on earth.
- Time and space are interdependent and form the fourth dimension, the *"time-space continuum"*, which can be understood in mathematical terms.
- The presence of matter results in the "warping or distorting" of the space-time continuum. This means that large bodies like a planet or star, with a strong gravitational field will create a curvature in the space around it.
- The force of gravity at the surface of the earth causes things (mass or matter) to accelerate at a rate of 32 feet per second per second. Acceleration due to gravity is

identical to acceleration due to a force pushing on an item.

We were created with five physical senses that were not designed to sense the relativeness of the universe. From the standpoint of the reception of the senses and our interpretation of those inputs, our environment is absolute. It is difficult for the average person to understand and visualize many of these concepts because they are outside the range of our normal senses. Yet, it is possible for humans to penetrate this barrier of understanding through training in mathematics and physics.

If you cannot understand this stuff, you are not alone, and don't worry about it because there are many others that do not understand it either. We do know that most of these concepts have been proven empirically through testing and experimentation. It is safe to have faith and believe in it. This beautiful universe we live in is not going to crumble because we do not understand. Leave it to the mathematicians and physicists to understand that stuff; that is their job.

Many people may tend to say, "Relativity is too complex and bizarre, it does not affect me, and I do not have to pay attention to it". Yet, relativity does play an important part in our everyday lives. Relativity affects us every moment of every day of our lives. We are the way we are because of relativity, yet most of us know so little about it. It is so commonplace in every aspect of our lives that we take it for granted and ignore it. We are simply unaware and oblivious to it. Before 1905, relativity was unknown to humans. The level of evolution of our Consciousness was not developed enough to know about it. At the turn of the last century, Einstein brought it to our attention, and now we can use it as an extra tool to help us continue our development. It is now important for us to know that it exists, and in general terms we should know how it works, because that will help us understand who we are, where we are going, and why we are the way we are. This book will attempt to give the basics of the commonplace manifestation of the law. Remember that from

Evolution of Consciousness

the standpoint of this book relativity is not a theory, it is real and it is a Universal Law.

5.1.4 Motion in the Astral and Spiritual Domains

Isaac Newton treated time, space, and motion as absolute. Einstein on the other hand discovered that Newton theory was not quite true; time, space and motion are relative. Einstein's relativity principles deal with relative motion in the physical universe only. There is no mention of astral motion or Spiritual motion. This book notes that there is motion in all domains of the universe. Physical and astral motions are closely related in that they both involve forms of matter, which have inertia. Spiritual motion is mental and has no inertia. Spiritual motion is also relative, but has a different degree of relativeness from that of physical motion.

5.1.5 Einstein's Mental Exercise (The Imaginary Train)

Einstein used imaginary mental thought exercises to help him see complex phenomena in his mind's eye. One such thought exercise that helped Einstein in developing the *special theory of relativity*, deals with straight-line uniform motion; it goes like this.

Consider four items involved in this mental exercise as follows.

(1) A long **train** traveling at a high rate of speed (close to the speed of light) on a straight railroad track.
(2) A **stationary platform** in front of the train station.
(3) A person called **Observer A**, standing on the platform.
(4) A person called **Observer B**, standing on top of a boxcar that is in the exact center of the long train and moving at a high rate of speed with the train.

- **Suppose the following scenario occurs**:

As Observer B (on the train) travels to a point directly in front of Observer A (on the platform), two lightning strikes occur. One lightning strike hits the front of the train, and the other hits the rear of the train.

- **Resulting remarks of the two Observers**:

Observer A and Observer B discuss the occurrence of the lightning strikes at a later time. Observer A (on the platform) states that he saw the two lightning strikes occur at exactly the same time. Observer B (on the train) states that he saw one lightning strike hit the front of the train first, and then a short time later a lightning strike hit the rear of the train.

- **Conclusions**:

There seems to be a discrepancy. Who is Correct? Einstein concluded that they were both correct. The apparent discrepancy was as a result of the two Observers being in different frames of reference. Observer B was moving (at close to the speed of light) with respect to Observer A.

- **Observer A reasoning (on the platform):**

Because light has a constant velocity the light from the front of the train and the light from the back of the train took the same amount of time to reach Observer A, therefore the two lightning strikes occurred simultaneously from the standpoint of Observer A.

- **Observer B reasoning (on the train):**

Because Observer B was moving forward (towards the light at the front of the train), the light from the front of the train had to travel a shorter distance, and so occurred first. On the other hand, the light from the rear of the train had to travel a longer distance to catch up, and so occurred later. Therefore, from the standpoint of Observer B, the two lightning strikes occurred at different times.

- **Relativity principle**:

Both observers were correct. This is so because observers in two different frames of reference will form different interpretations of the same event when speeds close to the speed of light are involved.

5.2 Relativity in Ordinary Things

If relativity is so prevalent in everything in our lives and in our universe, why are we so unaware of it? The main reason we are so unaware is that we were designed with five senses that interpret

Evolution of Consciousness

the physical world as absolute. Mother Nature did not see any need to encumber us with awareness of relativity when it was not going to contribute to our early growth. She preferred that we think of ourselves as absolute beings, living in an absolute world. But that was back when we were ignorant of our environment. We have progressed in our development to the point where we are now, and it is now practical to become aware of relativity in our lives. Awareness of relativity in our everyday live can help answer many questions that were unanswerable in the past. Besides, relativity is real, we live in a "relative reality", and we should know about it. We should also be aware that, over time, knowing the truth about relativity will have an affect in our thinking and subsequent philosophy of life. Another way to say it is that we have not been aware of relativity because our Consciousness was not advanced enough to understand it. At this point in our history Consciousness is reaching the level of advancement where it is practical to become fully aware of relativity. Understanding relativity now will help us accelerate our future development.

Several examples of everyday occurrences are provided here to try to get the idea of relativity across. We will start with things that we are familiar with and then advance to a treatment of such things as truth, happiness, good and bad, relative reality, and others.

5.2.1 Money is Relative

There is nothing fixed about money, it constantly changes value. The price of goods is always going up and down, mostly up it seems. Supply and demand is the main driving force in determining the value of money. Actually there are many factors that contribute to the constant changing in the value of money and all of them are relative. It is easy to see the price of gasoline change regularly at the pump. These pricing is actually very complex and involves global factors. It involves the crude oil production (OPEC for instance), transportation, refining, distribution, retail sales and many others factors in-between. Of course, all these factors are also relative. For instance, Supply and demand is variable and dependent on such things as the strength of the economy, the strength of the dollar, the whims of people, and many other factors.

The value of currency in the Foreign Exchange Market is extremely volatile. The exchange rate of one country's currency varies minute by minute relative to the exchange rate of another country's currency. Currencies are traded every day on the Foreign Exchange Floor. For instance trades are constantly being made between the US Dollar and the British Pound, the Japanese Yen and the European Euro, the Mexican Peso and the Canadian Dollar, and so it goes for all the currencies of the world. These exchange rates change in accordance with the relative strength of the currency in the various regions of the world. Rates change in relation to international demand and the strength of the economy in the various countries. One example is that if Americans buy more Japanese goods, then the demand for yen increases and the yen rises with respect to the dollar. This system is known as the *floating exchange rate*, which indicates that money is relative and there is no absolute reference point.

5.2.2 Relativity of Words

Many words in our language have evolved to convey the sense of relativeness in personal communications. The function of most adjectives is to describe or modify a noun. They also give significance as to the relativeness of the noun's function. For instance, the adjectives fast, faster, fastest can be used to depict the relative speed of an object or person. Other words that are used to indicate relativeness are such adverbs as fairly as in, "the turtle is fairly slow." or extremely as in "The acrobat is extremely agile". The same can be said of the adjectives slow, slower, slowest, or tall, taller, tallest, or fat, fatter, fattest and so on. These words are used to try to pin point the relativeness of the noun or actions. Beauty is relative to the interpretation of the observer thus the common expression, "Beauty is in the eyes of the beholder". That is, beauty is subjective and dependent on the interpretation of the viewer. There is no fixed reference point from which to describe beauty. Some may say that Miss Universe could be used as a fixed point of reference to judge female beauty, but even that changes from year to year according to the interpretations and feelings of the judges. Other words that are relative since their meaning changes with time are happiness, joy, fear, courage, and many more. The adverb relatively is widely used by many to indicate comparative relativeness of things. Some examples of these are relatively large, relatively loud, relatively cool, relatively fast, and so on.

5.2.3 Relative Humidity

Relative humidity represents the amount of moisture (water vapor) suspended in the air at a given temperature, relative to the total amount of moisture that the air can hold at that given temperature. Humidity is depicted in terms of a ratio or percent. The hotter the temperature of the air the more moisture it can hold, and conversely the colder the air the less moisture it can hold. There are many other words or phrases used to depict relativeness of things, such as relative density, relative quantity, relative pitch, relative speed, relative beauty, and so on.

5.2.4 Imaginary Political Spectrum

Politics is relative because ideology is relative, and ideology is relative since Human Consciousness is constantly evolving. In the United States we have two major political parties, the Republican and the Democratic parties. Each party has its own general political philosophy or ideology. An imaginary line called the *"political spectrum"* can be used to depict the spread of political ideology. It depicts political ideology as a subjective scale that varies in degrees of constituent opinions. The political spectrum (also called political opinion line or political scale) is completely relative. This political spectrum can be visualized as a straight line that has a center, a right (wing) and a left (wing). Arbitrarily we designate the Republican Party to be on the right of the political spectrum, and the Democratic Party to be on the left of the political spectrum. Each side is then subdivided and assigned names, to denote relational degrees of political philosophy to right or to the left of center. The Republican ideologies can be depicted by the following subjective terms starting at the center and extending to the right of the imaginary spectrum.

- Moderate or centrist Republican
- Fiscal Conservative
- Conservative
- Religious Conservative or Religious Right
- Social Conservative
- Ultra Conservative
- Far right wing conservative
- Right wing extremist.
- The most extreme ideology on the right is the bygone Nazism.

Other related ideologies of the right wing are the Libertarian Party, and the White Supremacists.

The Democratic side of the line, from the center and moving left, has general naming conventions such as:

- Moderate or centrist Democrat
- Liberal
- Fiscally Liberal.
- Socially liberal.
- Left wing liberal.
- Far left wing liberal.
- Left wing fanatic or extremist.
- The most extreme ideology on the left is the bygone Communism.

Other related ideologies of the left wing are the Green Party, the Labor Unions, the Black Caucus and many more. This list is only for illustration purposes and not intended to be the final word in politics. There are many experts in politics that have much better terminologies than these.

There are other political parties and many special interest groups in the United States whose political ideologies can also fit to the right or left of the same imaginary line, or they may even have their own political spectrum. Similar situations apply to politics all over the world; each country or group has their own set of political philosophies, agendas, and naming conventions. All of which are relative to the people that promulgate them.

Politics is always in a constant state of change; everything in it changes all the time. That is, the center of the spectrum may shift to the right or left, and the arbitrarily naming conventions also change. The relative ideologies are so fluid that they change significantly from one election period to the next. The positioning of each party as a whole, within the political spectrum, will shift according to the average ideology of its members. There is no absolute point of reference for the political spectrum; the positioning of each ideology is relative to each other. The definition of this imaginary political spectrum changes according to the average whim of its members. Each person generates its own ideology, which is a variable because the Consciousness of each individual changes as the individual develops and evolves.

By this time the reader may be saying, "So everything in our universe is relative, so what? That does not seem to be a big deal for me. It does not seem to affect me." Well, that is actually as it should be. We were designed to think we live in an absolute world while we are developing. But, as we develop, we finally are reaching a point where we want to know more of who we are, and we start to yearn for more knowledge. We are the way we are in large measure because of relativity.

5.3 Truth

We tend to think that truth, as we know it, is absolute. In the legal system it is common to hear words like, "Do you swear to tell the truth, the whole truth, and nothing but the truth, so help you God?" This is done in the trust that there is absolute truth to the matter at hand, and when a person knows the truth it is possible to know the whole truth. From the standpoint of relativity it is impossible to know the whole truth. The term at the end of the phrase "so help you God" could actually mean "the truth to the best of your ability". Truth like everything else in our universe is relative. Only God the Absolute knows the absolute truth. *From* our prospective and at our level of advancement, it is not possible for us to know the whole truth. At our level of development, most of what we know as truth is only partially true. Absolute truth never changes, but because we are in an evolutionary mode, truth does change for us. Several examples are included here to illustrate how truth has changed though history.

Prior to the Renaissance in Europe it was an accepted fact that the earth was flat. It took the discoveries of many Renaissance thinkers to establish that the earth was not flat that indeed it was a globe. Among the notables that contributed to the knowledge that the earth was round and that it orbits the sun along with the other planets were Copernicus, Columbus, Gilbert, Kepler, Galileo, Descartes, and Newton.

Our physical senses reported to our intellect that the earth was flat and that is what we believe to be the truth. After Columbus' discovery of the new land (1492), the general populace gradually began to realize that the earth was indeed a globe. Note that absolute truth did not change; it was only our interpretation of the truth that changed. Mother Nature knew all along that the earth was round. Learning this truth was a step forward in our progress towards Evolution of Consciousness.

Copernicus first alludes to the fact that the earth rotates on its axis and also revolves around the sun in 1540 when he published *On the Revolutions of Heavenly Orbs*. Galileo (1564 – 1642) demonstrated with the use of an optical telescope that the earth and other planets orbit around the sun. This new idea was unacceptable to the all-powerful church oligarchy of the day. The church preferred to think that the earth was the center of the universe since that is where humans live. Because of ignorance the church prosecuted Galileo for his discoveries. It took many years for the church to come to grips with the error of their assumptions and finally accept the truths as presented by Galileo. Again note that absolute truth did not change, only our interpretation of the truth changed.

Evolution of Consciousness

The scientific principles, as introduced by Aristotle, dominated ancient civilizations for around 2000 years. According to Aristotle (384 – 322 BC) the universe was made of "earth, water, air and fire". He did not know of gravity but he explained that all things tend to move in a downward direction because down is the natural state of rest for solids and liquids. Rocks for instance fall down, and water seeks and flows to the lowest point. Fire tends to go up and air goes up with it, but air tends to settle back down next to earth and water. He also thought that a rock would fall faster than a feather.

It was not until the seventeenth century, two thousand years later, that Galileo discovered that the pull of gravity is the same on all bodies and they all fall at the same rate in a vacuum. The vacuum is only to remove the affects of air resistance on the falling bodies. Isaac Newton (1642 – 1727) then discovered that mass is pulled by gravity and falls towards the center of the earth at a rate of acceleration of 32 ft/sec^2. Albert Einstein then eliminated gravity altogether and replaced it with the geometry of space-time which gives space a curvature. This stuff becomes difficult to grasp, but the reader can see that the truth according to what we understand is always changing. However, make sure to keep it in mind that Mother Nature's absolute truth never changes.

Only a few years ago the general public thought that the African Mountain Gorilla was a terrifying creature. It was depicted by fiction writers of the day as a vicious creature that was always ready to attack and rip us apart with its brute strength. Modern naturalists in recent years have turned that around by showing us that mountain gorillas are docile and intelligent animals that only want to be left alone. But instead, there are humans that are ruthlessly encroaching on their habitat, with the obvious danger of driving them to extinction. Similar scenarios can be said about snakes, sharks, alligators, and other wonderful creatures of our earth. In all these cases we interpret things out of ignorance, and every time we discover a new truth we remove a "veil of ignorance" and replace it with a "veil of knowledge". There are many myths in our society that promulgate untruths. One of the most prevalent is belief in the devil. Some people believe that there is a devil that is evil with great power and is continually battling with God. The concept of the devil is the only way these people can explain the source of evil. This is a myth that is caused by ignorance and will eventually be dissolved when those people advance sufficiently to realize the folly of their erroneous beliefs. Myths also abound in such things as fortune telling, faith healing, astrological predictions, palm reading, spiritual channel readings, and many others.

There are untruths in every walk of life. They are present in business, in the professions, in science, in medicine, in politics, in government, and in everything you can think of. If we knew the whole truth we would not be here. Instead we would be in a much higher level of evolution. It is normal to be ignorant of things; we are here to clear up ignorance and replace it with knowledge and truth.

Truth, in the relative reality that we live in, can only be partially true since we cannot know the absolute truth at our stage of development, but be comforted in the fact that truth as we know it is OK. We are to live our lives based on what we know to be the latest and best truth, and we should be ready to change as we discover new truths. When we discover a new truth, we should embrace it, and shift the old truth to the background as we would an old pair of worn out shoes. It is normal for us to not know everything. Do not fret if you do not know the whole truth. Also know that as you grow in understanding your definition of what you think is truth will change. You must be open to new understandings of truth and be willing to change because that represents growth.

Every time you discover a new truth it means that you removed a "veil of ignorance" and you move up another notch (level) in your evolutionary journey. Jesus said, *"Know the truth and the truth shall set you free"*. That is a true statement, but what this statement does not say is that it takes learning thousands upon thousands of little truths, one step at a time, to set you free, one step at a time.

It is common for older folks to reminisce of the times when they were young and reflect on differences of viewpoints between then and now. Those changes of viewpoints can be attributed to growth in understanding due to the experiences of life. It means discovering many new truths along the way.

5.3.1 Wealth and Happiness

We mentioned in the previous chapter that wealth is a relative quantity and while it is good in its proper place we should not grow attached to it. Happiness is also a relative thing and it is a common practice to think of happiness as emotionally connected to wealth or material things. All material things are relative and therefore temporary and if we depend on them for our happiness we will be disappointed when they are gone. When we look to the outside material world for our happiness; we will usually be disappointed. Material wealth is good because it does give us a degree of comfort and joy, it can give us a sense of pride and self-esteem, all of which is good, and it is good as long as we realize that it is temporary and therefore we should not grow too attached to it. The Consciousness can only find true happiness when it looks to the inside where its True Self resides. The only true source of happiness is inside, because that is where the absolute Human Spirit and Soul reside.

5.4 Reality

We are inclined to think that real things are tangible things that we can see, feel, touch, hear, and taste. We think of chairs, cars, mountains, people etc. as being real. It is normal for us to perceive our reality as real and absolute. We were designed to accept our environment as real and absolute and that is as it should be. It turns out that all these things that we call real are only partially real. What we call physical reality is relative and is partially an illusion. It is difficult for most people to accept this concept. It should be understood however, that the perception of absoluteness is only a perception. The world appears real to us, but it is only an apparent realness.

It is well known and accepted that our perception of reality has changed drastically over the last one hundred years due to the innovations and advancements in science and technology. Our perception of the size of the world has changed over the past few centuries. For instance, two thousand years ago the earth was perceived as infinitely large, by the start of the 19th century the world was perceived as extremely large, and by the start of the 21st century the earth is now relatively small. Our perception of the size of the world has changed partly because of changes in technology, but most importantly because our advances in conscious awareness. At the beginning of the 20th century high tech consisted of the telegraph and the railroad track with the steam engine. The automobile with the gasoline engine was in its infancy. Most of the transportation was still by horse and buggy. One hundred years ago a 100-mile trip by horse and buggy was normally a three-day trip, which was a major undertaking. Today, at the beginning of the 21st century, it is no big deal to hop into a modern automobile and travel 100 miles on excellent freeways in less than 2 hours. Many advances in science and technology have profoundly changed our perception of the world. Some of the recent technologies that have contributed to our fast change in perception are such things as the automobile, airplane, communication satellites, information technology, radio, television, the Internet, and many others.

5.4.1 The Affects of Modern Physics on the Perception of Reality

Many discoveries of modern physics have changed the way we think. Discoveries in physics have removed old myths and expanded our perception of reality, yet we still have a long way to go. Einstein's *Theory of Relativity* and the discovery of Quantum Mechanics have had profound affects in the development of our new technologies, and have also indirectly changed our perspective of reality. Quantum Mechanics has described the most basic and fundamental building blocks of our universe as vibrations, which can be expressed in terms of mathematical equations. Chapter 3 dealt with waves and established that the universe is made of waves. In our discussion of Quantum Mechanics we discussed that Quantum Mechanics is expressed by a set of mathematical equations that defines all matter in terms of wave theory. We also discussed that the

Evolution of Consciousness

microrealm is made of atoms, molecules, and subatomic particles, which in turn are made of nonphysical waves. So, it seems that what we normally call real or reality is made of vibrating waves.

An imaginary story provided as an illustration is as follows: Suppose that we could change the scale of the atom, and we equate the proton of a hydrogen atom to be the size of a basketball and the electron to be the size of pinhead. If such equivalence existed then the pinhead would revolve around the basketball at a distance of fifteen miles and nothing but empty space would exist between the basketball and the pinhead. This little story suggests that matter is made of mostly empty space with a tiny amount of wave substance within that space.

If you slam your fist on the surface of a hardwood table you can immediately conclude that the table is solid. Normally you do not think that both the table and the fist are made of atoms and molecules made mostly of empty space that has electromagnetic properties that repel each other. Even though the molecules of the table and fist are mostly empty space the fist does not go through the table due to the molecular repulsive forces. The perception of our senses is not that the empty space of the fist did not go through the empty space of the table. The perception is that the table is solid and hard. Our senses were designed that way for a reason and we should continue to abide by the perception of the senses.

5.4.2 Relative Reality

Based on the preceding discussion our reality is by no means an *"absolute reality"*. This book asserts that we live in a *"relative reality"*, but this relative reality is real for us at our present level of development and that is as it should be. We should also be aware that because our reality is relative, it would continue to gradually change as we advance forward in our ongoing evolutionary progress.

5.5 Good, Bad, Right, and Wrong?

Humans tend to struggle with the definition of good and bad, or right and wrong. People often ask the question, "What is good?"

or "What is right?" or "What should I use as my guide to live a virtuous and moral life?" The definitions of good and bad seem to fluctuate from culture to culture, or from one part of the world to another, or with different degrees of education, or with different age groups, or with different ages of our history and so on. The definitions are always changing, always being modified, or improved, or rejected, and they also vary in degrees of goodness or badness from person to person. More often than not, many things that were considered good at one time changed to bad, and other things changed from bad to good. Many changes occur as the person matures or develops intellectually. The reason the definition changes so much is that good and bad, right and wrong, virtuosity, morality, and right living, are all relative things. As we remove the "veils of ignorance" and replace them with "veils of knowledge", we change our definition.

It is important that we obtain the best possible definition of good and bad since we are philosophically bound by whatever definition we have at any particular time. Those definitions guide our lives, but they are only good until we advance another notch along the ladder of evolution, and then we change our definitions again. This process goes on and on, throughout our evolutionary journey and it is normal. Mother Nature does not change her definitions of good and bad, only evolving humans change. Why the big discrepancy? The discrepancy is that humans are relative and evolving whereas Mother Nature is absolute. A few examples of things that have changed from good to bad and vice versa are given here to try to clarify the point.

- *Slavery*: At one time in American history, slavery was considered "right" by slave owners of the South. Slavery was essential for the well being of the economy of the south. It took a civil war to abolish slavery in the United States. Nowadays slavery is universally considered "wrong". Slavery is man-made and produced in part because of greed, and disregard for human sacredness, but above all it is a product of ignorance. It must be remembered that our ignorance is normal and natural, and

Evolution of Consciousness

will be corrected as the society evolves, after much pain and suffering.
- *Whaling:* People of the 1800s used to think that killing whales was "good" and proper. Whaling provided exotic oils and wealth to the whaling community, and besides, whales were animals that were inferior and subject to our exploitation. Nowadays most of us think it is "evil, cruel and ruthless" to kill whales for profit. We now think of whales as huge and intelligent wonders of nature and should be protected.
- *Predators:* Wolves and mountain lions are predators that were practically decimated in the American West during the late 19th century and early 20th century. Predators were considered destructive of livestock and dangerous to humans. Early pioneers had an inborn fear and hatred for wolves and mountain lions. Pioneers considered predator's fair game and were no match to their rifles, consequently they were almost completely wiped out. In recent years our naturalists have realized the folly of killing those natural creatures and have begun to introduce them back into the wild, in such places as Yellowstone National Park. Naturalists now know that predators and prey form a natural balance and are good for each other. The predator only kills to eat, and they only kill the old or sick prey, which has the net affect of keeping a healthy balance in the wild. It is now known that everything in nature is connected to everything else. What the pioneers of old considered "bad", are now considered "good" by our modern naturalist and society as a whole.
- *Pollution*: In the early part of our industrial revolution (19th and 20th centuries) we used to think it was OK to dump our waste products into rivers, lakes, and into the atmosphere. That has changed and it is no longer OK to pollute. We are taking big steps to raise the awareness of the population in general and to motivate the political will to act.

- *Wars*: Wars throughout history have been good for the winner, and bad for the loser. It has always been OK for the stronger warring party to kill, plunder, and steal from their enemy. Normally the winner takes great pride and shows great joyful exuberance after the enemy has been ruthlessly conquered. On the other hand, the vanquished suffers pain and humiliation, and usually vows to take revenge that is accompanied with great hate. War and destruction are still common in our world today, but there is a sense among many thoughtful people that war and violence are wrong, because they cause a lot of needless atrocities, pain, suffering, agony, and despair. In fact both antagonists always loose. Wars are man-made, and create havoc for many. A common misgiving that some people have during a time of peril is that they lose their faith in God, and they may exclaim, "Where is God in this time of terror? Or "How can God allow these perilous things to happen?" From God's point of view, man-made wars and terrorization are strictly artificially created by humans, and are due to the greed and ignorance of humans at our level of Consciousness. However, they are within the boundaries of Universal Law and serve as horrible lessons to the contestants. God does not create wars, and neither does He suffer pain or human loss. What appears to us as material destruction, pain, suffering, and horrible death is relative and applies to humans only. God never loses material things, and never loses human lives to so-called death. Human death only applies to the physical body; the Real Self never dies it just moves on to come back another day.
- *Weather*: We often think of weather as being good or bad. We think of good weather as being pleasant to us. For example, good weather is sunny, cool, with a light breeze blowing. We think of bad weather as being unpleasant to us. Bad weather would be a cold blizzard with rain and snow streaking horizontally. Of course our defini-

tion of good and bad weather is only relative to us. To God the Absolute all weather is "good", and there is no such thing as "bad" weather. All weather is created for a good reason.
- *Natural disasters*: We always consider natural disasters such as hurricanes, tornadoes, floods, earthquakes, and volcanoes as "bad", and rightfully so. But, they are "bad", only relative to us. Relative to God the Absolute, all natural disasters are good, and were created for a reason. They have been occurring since before humans first set foot on the earth and are part of God's grand design of an ever-changing world.
- Homosexuality and same sex unions have been highly controversial in recent years in the United States. Recent polls indicate that the average public or political opinion has been changing to become more tolerant of same sex unions. This is just another example of the overall Consciousness of the public gradually changing as it advances forward.
- There are many examples of church traditions changing and causing the rules to be altered to accommodate the shifting Consciousness of its parishioners. For example it used to be a sin to eat red meat on Fridays, because the Catholic Church advocated eating fish. That changed when it became obvious to the average that it was a silly rule that was not based on any significant Spiritual reason. There are similar myths about eating pork or eating beef in other religious orders throughout the world.

5.5.1 A Good Universe

Good and bad, right and wrong apply to humans only since we are relative beings. God is "good" and He designed a good and perfect universe. From the standpoint of God the Absolute, there is no such thing as "bad" in our universe. The whole Universal System design including the Universal Laws, mind, energy, matter, life and Consciousness are good. You may ask, "If this is so good, than why is it that we have so much bad happening in our world and in our lives? Why is it that we have wars, atrocities, terrorism, poverty, hunger, pestilence, and the like throughout the world?" The reason is that bad is man-made, and it is bad relative to us only. God does not create "bad". It is because of our ignorance that we create bad things and therefore we cause our own suffering. The "Law of Cause and Effect" will not allow us to go scot-free for our misdeeds. Because of this law we eventually learn the folly of our behavior and eventually learn to mend our wrongful conduct.

It is important that we develop the best possible conception of right and wrong, good and bad, morality, virtuosity, and right living. But, if all these things are relative, and always changing, how are we to know what the true definition of good is at any given time? In actuality, we don't know! It is definitely a dilemma of major proportions! It is a paradox of nature. Good and bad are relative things and our definitions changes as we evolve and get wiser. In the mean time, if we are intent in living virtuous and righteous lives, what are we to do? How should we live our lives, if we do not know the true definition of right and wrong? A partial answer goes like this: We cannot know the "absolute good" since only God knows that, but we do not have to know the exact definition of good and bad. We only have to live our lives to the best of our ability. ***Our best good at any given time is to take the course of action that is best suited to the requirements of our individual Consciousness in the next higher step of our development***. How do we know what is best suited for our development at any given time. The best course of action is always that which helps us advance in Consciousness, and if a course of action detracts from our development then it is bad. Because of relativity there is no clear, concise, and straightforward answer. The answer is far to complex for us to understand, so it is best to go back to the previously mentioned "game of life, or Divine Plan" where we do the best we know how and leave it up to the Divine Mind to take its course in our behalf.

Each one of us is at a different level of evolution and the best good for each person varies. It is natural for our definition of right

Evolution of Consciousness

and wrong, good and bad to change as we develop. It is a significant part of our evolutionary process. The next section is an effort to try to shed more light on the subject of what is the best good for us as we move forward.

5.6 Ethics and Rules of Conduct

Many philosophers have struggled throughout history to define "ethics". Many sources of deep thought have been used to get answers to the dilemma of, "What is good, bad, right and wrong?" "What is ethical, and how do we know what the rules of conduct are?" There is no easy way to define ethics or rules of conduct because they are relative and it is like trying to hit a moving target. Because they are relative they keep changing as we change. Many philosophies have been developed throughout the ages by the best thinkers of the day. Most of the time these philosophies are OK for that society at its particular period in history, but as the society evolves the ethical standards change. Many of the previous philosophies of ethics and rules of conduct seem to have merit but they also have faults. The reason that philosophers of the past failed to arrive at a true answer to the question of ethics is that there is no true answer that applies to all ages. Philosophers have always failed to take into account that we are relative beings living in a relative universe and as a result, our concept of good and bad changes. Philosophers always tried to define ethics in what they believed to be absolute terms and for the most part those terms were OK for the specific period of time. But, with passage of time their original definitions did not apply and new ethics definitions had to be developed. It is impossible to define relative ethics using relative ideas. We cannot set one firm collection of ethical rules that will apply to all times; we are just not smart enough to do that. Human Consciousnesses is always in a state of unfoldment, always moving from lower to higher, and this unfoldment is the aim of life; it is the Divine plan. We do not stand still for the ethics of the past, we have to move on and the ethics have to continually try to catch up with us.

Philosophers have used many sources to develop their ethical definitions of the past. Three foundations that seem to stand out and have been used to try to make sense of the rules of ethics and conduct are as follows:
(1) The Spiritual or Religious Revelations
(2) Intuition
(3) Human Reason and Human Made Laws
The merits and faults of each will now be address.

5.6.1 Spiritual or Religious Revelations

Every society throughout history has had religious beliefs that are based on some insight that was delivered by some advanced persons. These advanced persons can be thought of as highly advanced Consciousnesses that came from higher realms of the universe to teach in a particular time in history. This book designates those higher realms as Spiritual Domain 2 (SD2) and/or Spiritual Domain 3 (SD3). These teachings are meant to meet the basic Spiritual or ethical needs of the society at that particular time in their development. People with special insight are called Sages, Prophets, Priests or Teachers. The best known of these Sages include Moses, Jesus, Mohammad, and Buddha. Each provided the foundation for a major religion. These Teachers revealed many Divine truths that provided guidelines for their followers to live by. The older original teachings go back to antiquity and some of those old teachings may still partially apply today but most can be considered obsolete for our modern society. For example, the "Ten Commandments" of Moses were introduced to humanity (through Moses and the Jewish people) about four thousand years ago. The Commandments were perfect for the people of that time and may have been somewhat advanced for them. But, surely we have advanced a great deal in the last four thousand years and even though the Ten Commandments do apply today they apply in limited degrees. There are much better definitions of right and wrong today and should supersede the Ten Commandments.

It must be remembered that the teachings were delivered to meet the needs of the people of the day. The teachings normally were composed of Divine truths but were sculptured so that the people of a lower level of advancement could at least understand the general principles. Most of the time the teacher did not write down the details of their ministry and so much was lost. The teachings of Jesus were true and correct and he used parables to get his point across to the people of the time. Jesus did not write down his teachings and a lot was lost before other writers finally wrote down the gist of his teachings. Another thing that happens is that over the millennium of time many of the original teachings are distorted and/or magnified, or convoluted by legend, superstition, and myth. Many of the old teachings were provided for people of the ancient world and are difficult for people of today to understand the symbolism. In essence these teachings cannot be used as infallible rules of conduct for all time.

The supporters of the old religious philosophies are usually priests and people of deep religious convictions. Many espouse that all the rules for living can be found in these scriptures because the scriptures are divinely inspired. Opponents argue that though many of the old scriptures were divinely inspired they were provided for ancient people who at the time were not very advanced intellectually. Therefore these writings do not apply for people of today who have advanced a great deal. People of today are always ready for new revelations that apply to their present level of development.

There have been many people throughout the ages who have imparted knowledge and inspiration to us. In modern times we have Albert Einstein, Linus Paling, Mahatma Gandhi, Mother Teresa, Martin Luther King, Tomas Edison, and many others who have provided us with wonderful insights of truth, knowledge, and goodness. The wonderful accomplishments, deeds, and revelations of many of our contemporaries have been exemplary, but the rule is that other great men and women will sooner or later surpass those good deeds and examples.

5.6.2 Intuition

The second foundation for ethics is intuition, which comes to us through the Intuitive Mind. The Intuitive Mind is the higher subconscious mind, which has a direct connection to the Higher Self (Human Soul and Spirit) and to other Higher Intelligences. The Intuitive Mind is not a thinking mind it is a knowing mind, so that the information coming from this foundation is good and is called "inspiration". We should learn to cultivate this gift. The problem with intuition is that everybody is at a different level of development and consequently everyone receives different results. Furthermore, the less developed person tends to use this gift to a lesser degree while the more developed person tends to use it to a greater degree. To explain the workings of this principle refer to the section on the "veils of ignorance" in Chapter 4. In this analogy the light at the center represents the wisdom of the Spirit, the bulb represents the Intuitive Mind, and the "veils of ignorance" represents the person with all his or her individual obstructions. The person that has the full compliment of the "veils of ignorance" is a person of low development. A more advanced person has removed many of the outer "veils of ignorance" and has replaced them with "veils of knowledge". Therefore the person of low development cannot get the light of the Spirit because there is too much material that smothers the light. The advanced person has less of the veils therefore less material and the light can penetrate to the outer veils. Because every person is at a different level on the scale of evolution, every person gets a different prospective of the inspiration of the Spirit. This is the reason that everyone who uses intuition comes up with different answers.

The fact that no two people are exactly alike implies that there are as many standards of morality and conduct, as there are people. If everyone went around saying, "My conscience approves of it therefore I can do whatever I want." then we would probably have anarchy on our hands. The insight received by one person does not necessarily match the insight received by the next person. So, this would not apply to defining what is good for the society. Nevertheless, learning to listen to the inner self is good practice. Generally the way this works is that when you make a choice to do something you should listen to your feelings; if you get a good feeling of comfort and joy, you are probably making the right choice; if you get a feeling of discomfort, or uneasiness, or regret, you are probably going down the wrong path. Either way you have a choice; if you made the right choice you will be rewarded, and if you make the wrong choice you will suffer anguish, which is proportional to the infraction. This is so of course because of the Law of Cause and Effect.

5.6.3 Human Reason and Human Made Laws

The third source of ethics comes from human reason. Humans are social creatures and naturally tend to organize into groups, tribes, or social organizations. Human societies tend to develop rules or laws to govern themselves by. In a democracy, governing bodies are elected by the people to represent the will of the general populace, and to create laws to govern the constituents. Human made laws generally represent the average will of the people governed. Laws have been evolving through the history of humanity and are essential to a civilized society. Today governmental laws have reached a high degree of sophistication and represent the best that the society as a whole symbolizes. There is always a struggle to provide maximum order without eroding the freedom of the people. These laws are generally good, they have been evolving over many years, and they are relative in nature, therefore they will continue to change. As the society evolves forward, the laws also evolve to keep up with the development of the society. Human made laws tend to lag a little behind the median level of the people because a certain amount of political will is required to overcome the momentum of the masses. There are always some stragglers in the society that tend to slow down progress.

Man made Laws, bylaws, rules and regulations are based on human reason that is based on human experience gained over ages of time. They generally represent the average level of development of the people being governed by those laws. In the United States this applies at all levels of government to include the federal, state, and local governments. Eventually laws become old and obsolete and no longer satisfy the requirements of the governed. Old laws are then subject to be replaced with new laws that are more closely aligned to the level of Consciousness of the constituency.

5.6.4 Right Action

Each one of the three foundations of ethics or right conduct that have been presented has merits as well as faults. The rule is that each person should live his or her life to the best of their ability, using the influence of "revelation, intuition, and reason". All three foundations affect each person in varying degrees and the "best good" for a person is probably a composite of the three foundations. "Right action" is whatever best suits the requirements of the individual Consciousness, or whatever it is that helps that individual Consciousness in its next higher step of development. "Right Action" is the highest course of action for the development of the individual Consciousness at a particular time in its evolution. Right and wrong, good and bad are relative terms. A "relative wrong" should be classified as merely a low conception of "right", or an action falling short of complying with the highest conception of "right".

The primary purpose for Consciousnesses is to evolve and unfold. Whatever action we do that helps that unfoldment is "good or right" whatever actions we do that detracts from our unfoldment is "bad or wrong". The "right or good" falls in line with the plan of "Evolution of Consciousness" while "wrong or bad" detracts from the plan. What is "right and good" varies from person to person according to their level of unfoldment. An extreme example is that it is "right" for a tiger to be ruthless and savagely kill prey for food, because that is in accordance to its level of development, but it is "wrong" for a developed human to revert to that stage. For an advanced human to harbor feelings of hate, revenge, jealousy and the like would be "wrong" for it would mean going back to levels of development long past, and it would be contrary to the reason and intuition of the person. I hope this gives the reader an idea of the complexities of right and wrong, good and bad, and ethics in accordance with the Law of Relativity.

5.7 Relativeness of the Consciousness

Einstein's "Theory of Relativity" deals with relativity as it pertains to the inanimate universe and it is only apparent when dealing with speeds that are close to the speed of light. There is no single frame of reference in the inanimate universe; things are always defined or measured in relation to something else. In Einstein's relativity a set of coordinates have to be referenced with respect to some other set of coordinates. When it comes to the Consciousness, Einstein's relativity principles do not apply. The Consciousness is different from the inanimate universe. Each and every Consciousness is a master frame of reference. From the standpoint of the each individual Consciousness the universe is what it is, with respect to itself. Each Consciousness is its own reference point and the universe revolves around it. To complicate matters even more, each Consciousness is relative and by definition relative things change and evolve forward. So, as each Consciousness chances its own point of reference also changes. That is, its view of the inanimate universe changes accordingly. It is obvious that the Consciousness and universe combination is extremely complex. If it were possible to develop a mathematical equation to represent the operation of the Consciousness in the universe it would have to contain a huge amount of complex functions with each function having an infinite amount of variables. It is hard to imagine it. This book advances the following theorems:

1. Relativity and evolution are completely interrelated and complementary; you cannot have one without the other. That is, because the Consciousness is relative it evolves, and conversely because the Consciousness evolves it is relative. The same applies to the universe as a whole.
2. Truth, good, bad, right, wrong happiness, wealth, and everything else is relative because the Consciousness is relative. By the same token the universe is relative because the Consciousness is relative. Furthermore the universe is relative to the same degree as the Consciousness and as the Consciousness advances so does the universe. And to take it a step farther, the universe exits only because the Consciousness

exists. If there were no Consciousnesses, the existence of the universe would not be possible. The Consciousness is what gives expression to the universe and not the other way around.
3. The level of evolution of the universe is dependent on the level of evolution of each individual Consciousness, which would imply that there are as many universes as there are Consciousnesses.
4. The implication of these statements is profound; this means that the Consciousness is the most important thing in the universe and consequently the Consciousness is everything. If you say "I AM" and you think of yourself as being the universe you are correct since your universe exists only because you are a Consciousness.

As the Consciousness evolves forward, so do all other relative things evolve along with the Consciousness. The Consciousness evolves in degrees; it started out primitive and progresses one degree at a time to its ultimate level. Bear in mind that while things are changing from the viewpoint of the Consciousness, nothing ever changes from the viewpoint of Mother Nature since She is Absolute. This is a paradox that is difficult to understand, but perhaps it will become clearer with some amount of deep thought and contemplation.

5.7.1 Scales of Nature

There are many scales in nature and each scale represents a particular parameter or characteristic of nature. These scales are interdependent variables that change in degree and are a result of the changing Consciousness. Each is represented by its own range of vibrating frequencies and each range of frequencies has its own signature frequency that depicts its position along the scale at any particular time. This signature frequency is like a beacon to Mother Nature which points its relative position on the scale at all times. Four of these scales are as following:

1. Scale of Evolution of Consciousness – The Consciousness rides this scale from beginning to end. The scale starts at the most primitive level of human evolution (caveman) and proceeds upward through eons of time and through every stage of Consciousness until it reaches the highest level of human evolution. As the Consciousness evolves its signature frequency also increases in correspondence and in direct proportionality to the level of its evolution. The signature frequency depicts accurately the level of progress of the Consciousness; it pinpoints the location on the scale so that Mother Nature can determine precisely where on the scale the Consciousness is.
2. Scale of Ignorance – This scale changes in degrees from ignorant at the primitive level of Consciousness to intelligent at the completion of the evolutionary journey.
3. Scale of Relativity – This scale changes in degrees from very relative at the primitive level of Consciousness to absoluteness at the completion of the evolutionary journey.
4. Scales of Emotions – Each emotion has its own individual scale, which varies in degrees as the Consciousness evolves. The lowest end of the scale represents the particular emotion at the animalistic level and the highest end of the emotion represents the virtuous end of the same emotion. A signature frequency represents the exact position of the emotion on the scale at any particular time. Each Consciousness carries with it its own complete set of emotional scales throughout its entire journey.
5. Scale of Love – This scale varies from animalistic lust for sex at the bottom of the scale to Divine Love at the top of the scale.

5.7.2 Relativeness of the individual Human Components

The human being is composed of ten component parts and the relativeness of the individual components differ one from the other. As already mentioned relativity is not a constant function, it is a variable function that varies in degrees. There are no mathematical equations to quantify relativeness so we are confined to using comparative language to try to depict some semblance of value of relativity. The best we can do to give a semblance of quantitative associations is to use such words such as greater than, less than, proportional to, equal to, and so forth. Refer to the list below of human component parts. It is arranged in descending order from the most significant component part (Spirit, at the top of the list), to the least significant component part (physical body, at the bottom of the list). Included with the list is a classification of relativeness between components.

10. Spirit – The Spirit is absolute and is not affected by time and space.
9. Soul – The Soul is absolute and is not affected by time and space.
8. Consciousness – The Consciousness is of a variable relativeness. It starts out relative and becomes less relative as it evolves forward. It approaches absoluteness as it reaches the highest levels of development. Consciousness is aware of and is affected by time and space only during its lifetimes on earth in a human physical body, but when living its life in the Spiritual Domains it is oblivious of time and space. Overall life in the Spiritual Domains is more natural for the Consciousness. Earth lives are special conditions for the Consciousness, which are used for personal development at a particular stage of its growth. An analogy to this is going to school for a certain part of your early life but most of your normal life is spent working, earning a living, and raising a family.
7. Intuitive Mind – The Intuitive Mind is variable in proportion to the relativeness of the Consciousness and is not affected by time and space.
6. Intellectual Mind – The Intellectual Mind is a variable and is proportional to the relativeness of the Consciousness. It is highly affected by time and space when living an earth life.

5. Personality – The Personality is a variable and is proportional to the relativeness of the Consciousness and it is highly affected by time and space.
4. Human Instinctive Operating Mind (HIOM) – The HIOM is relative throughout its tenure. It is a subconscious mind and unaware of time but is affected by the rhythm of time. The HIOM is only active in the Physical Domain in living organisms.
3. Vital Energy – Vital Energy is relative and is affected by time and space.
2. Astral Body – The Astral Body is relativity and is highly affected by time and space.
1. Physical Body – The physical Body is relative and is highly affected by time and space.

When consciousness reaches the highest possible level of its evolution it becomes absolute and merges with the Soul and the Spirit. At that time the Spirit-Soul-Consciousness combination is ready to go home as a unit and become a part of God the Absolute. Components 1 through 7 become meaningless at that time and are completely discarded

5.8 Relativity of Domains

This book divides the universe into five different domains that exhibit marked differences of characteristics, and will be discussed in more detail in Chapters 7 and 8. As with everything in the universe, the domains are arranged in a hierarchical order. Note, we do not have multiple universes; we have only one universe with multiple domains. The relativeness of each domain is proportional to the average relativeness of the Consciousnesses that live in them. The Physical Domain, Astral Domain, and Spiritual Domain 1, are the most relative since people of the Human Consciousness Stage, which is the lowest stage of human development, occupy these domains. Spiritual Domain 2 is less relative and more absolute than Spiritual Domain 1 since the more advanced people of the Soul Consciousness Stage and Cosmic Consciousness Stage occupy it.

Spiritual Domain 3 is the least relative and hence approaches absoluteness since it is occupied by the highest stage of human development, which is the Spiritual Consciousness Stage.

Chapter 6

Evolution

Everything in the universe evolves from lesser to higher forms, and then from higher to even higher forms. This book considers evolution to be a Universal Law. It manifests itself in every component of our universe, to include the Physical Domain, the Astral Domain as well as the Spiritual Domains. Many principles of physical evolution are well known, and are generally well accepted by most segments of our society. A few religious segments do not accept evolution because it tends to conflict with their religious beliefs. Prior to Charles Darwin evolution was practically unknown to the population at large. Actually, Evolution was known to a few intellectuals since Aristotle stated that it is obvious that higher life forms came from lower forms. Darwin introduced us to evolution of living organisms in a purely physical world. We touch on the evolution of living organisms as well as evolution of the physical cosmos only to tie in to the greater topic: the Evolution of Consciousness as well as the evolution of the mind. Evolution of Consciousness takes place in the Physical as well as the Spiritual Domains, and it is vaguely understood and recognized in our society. Also discussed is the metaphysical explanation of evolution of the Physical Domain, which has many similarities to the scientific theories.

6.1 Cosmic Evolution

Cosmic evolution is also known as cosmology. It is the study of the origin, structure, and development of the physical universe or Physical Domain. Cosmology tries to explain how the universe began, how it developed to the present stage over billions of years, and what it will be like in the future. Scientists nowadays have new tools at their disposal and have made huge strides in the advancement of the science of cosmology. Scientists use the aid of mathematics, Einstein's Theory of Relativity, and Newton's law of gravitation to study and develop their theories on the evolution of the universe. In recent years scientists have had the advantage of new space exploration tools due to the advances in space technology, thanks to the National Aeronautics and Space Administration (NASA). Some of these tools include the Space Shuttle, the Hubble Space Telescope, and the International Space Station. There are also several new space satellites that take readings of the sky in the infrared, ultraviolet, x-ray, and high-energy gamma ray. There are terrestrial telescopes that also focus on these areas of the electromagnetic spectrum and have contributed enormously to recent developments. These tools have greatly enhanced and will continue to enhance the development of the science of cosmology and its evolution.

At present the latest scientific theory regarding the origin of the universe is explained by the theory of the "big bang", which has wide acceptance by the scientific community. Another theory that has gained recognition, and will be touched upon in this book is the theory of the "black holes" in space. These two phenomena, the "big bang" and the "black holes", are considered crucial in explaining the beginning and ending of cosmic evolution. The basic principles of both of these theories will be discussed from both the scientific and the metaphysical standpoints.

6.1.1 Origin of the Universe

The *"big bang"* theory is the most popular theory for explaining the origin of the universe. This theory basically states that the universe started with a huge explosion of *"something"* that was apparently infinitely compressed. There is no clue among scientists as to what this *"something"* was, but it is assumed that at time zero this *"something"* was triggered into a gigantic explosion. Scientists do not know what this mass was before time zero, but they can speculate as to what it was like millisecond after time zero. Using the Newton's law of gravitation, Einstein's Theory of Relativity, and heavy-duty mathematical exercises scientists feel that they can speculate within reason about the initial beginning.

Many scientists agree that the "big bang" occurred approximately 13.7 billion years ago. It is thought that originally the universe contained only the basic particles of matter such as electrons, protons, and neutrons. These particles eventually formed the lighter elements of matter such as hydrogen and helium. As the universe expanded and cooled, hydrogen begin to form clusters of gas, which eventually ignited and became stars like our Sun. In time, suns were grouped together to form galaxies of stars. The process inside the stars is a nuclear reaction, much like the hydrogen bomb that the U. S. Government and others have developed and tested. The nuclear reaction of the sun is called *"fusion"*. It involves the combining (fusing) of hydrogen atoms, which results in energy release, and the formation of helium atoms. Actually this process converts a small amount of matter into energy in accordance with Einstein's equation, $e = mc^2$. These nuclear reactions go on for millions of years until the hydrogen is exhausted. After the hydrogen depletes in the sun or star, the nuclear reaction continues using helium as its fuel, which in turn creates heavier element. These reactions continue for more millions of years, and the byproducts of the reactions are release of energy, and the formation of heavier forms of matter, such as, carbon, oxygen, nitrogen, iron, as well as all the other naturally occurring elements in our periodic table.

The next step in the evolution of stars is that after a few billion years, the stars grow old and start dying. As they die they leave behind all those heavier elements that were made during the various stages of the star's existence. The heavier elements eventually break apart from the original star mass and float off into space. Eventually some of these materials begin to collide with other materials in space and coalesce to form planets. That is, these space materials congeal and solidify into planets, and in the case of our own earth, it became part of a "Solar System". Our solar system consists of the sun and nine planets. Scientists do not explain how the laws of physics were developed except that they exist. Apparently the laws came about automatically as a result of the big bang. Scientific methods are totally objective; every hypothesis requires verification with test and experimentation. Science does not like subjectivity; consequently there is no reference or implication that God had anything to do with the big bang.

6.1.2 Contraction of the Universe (the Black Hole)

The principle of the *black hole* theory is that an intense gravitational field develops when a very large star, much larger than our sun, starts to decay and die due to consummation and depletion of its nuclear fuel. When the star exhausts its nuclear fuel it begins to cool; this cooling affect will eventually causes it to collapse. The collapse causes the material of the star to become very densely compacted into a relatively small volume, which causes gravity to increase immensely in that small volume. The strong force of gravity causes other material in the vicinity of space to be pulled into the black hole. As more and more matter comes crashing in, the gravitational force continues to increase, which in turn causes further compactness and compression. This is a feedback loop in which gravitation causes matter to come crashing, which causes higher density, which causes higher gravity. Eventually the compression becomes so great, and the gravitational attraction so strong that even light is sucked in, and once in, nothing can escape. The fact that light can only go in and not out gives it the name black hole. Black holes cannot be seen directly with a telescope because there is no light coming out for us to see. Scientists have to rely on other evidence to surmise that they exist. Astronomers have seen evidence of seven black holes in our Milky Way Galaxy. There are some astronomers who believe that there may be millions of black holes in our galaxy, which cannot be detected with our telescope. Our scientists regard black holes to oc-

Evolution of Consciousness

cur as a result of natural phenomenon and make no mention of God having any thing to do with their creation.

6.1.3 Evolution of the Earth

Like everything else in the universe, the earth had a beginning and has been changing and evolving ever since. Scientists estimate that the earth started to form about 4 1/2 billion years ago. They have several theories of how the Solar System and the earth formed in the beginning, but the theories are still vague due to the complexity of the matter. The most appealing theory, I believe, is one that states that the solar system developed from a spiraling *nebula*. A nebula is composed of pieces of rock, metal, dust, and clouds of gas. The sun probably formed at the center of the nebula, and the other debris flattens into a disk around the sun. The debris then clustered into spinning whirlpools, and over a few more million years formed into planets.

Scientific theories state that the material that accumulated to form the earth grew large enough to cause an increase in its gravitational force. Gravity in turn caused more material from space to be attracted, accumulated and compressed. The compressed material became hot, melted and formed the molten core of the center of the planet. Heavy metals sank to the center of the earth and the lighter material and gases went to the surface. Lighter materials eventually formed the crust, the water, and the atmosphere of the earth. Over millions of years the water accumulated in the low areas forming the oceans; the higher areas of land formed continents. The earth's early atmosphere probably contained large amounts of carbon dioxide, similar to present day planet Venus. Oxygen was probably not abundant in the atmosphere until plants started to appear on the earth.

The earth has been changing from the very beginning and continues to change to this day. Events that cause change on earth are:

- Plate tectonics which cause continental drift and sea-floor spreading

- Earthquakes
- Volcanoes
- Weathering and erosion

These events create mountains, lakes, canyon, rivers, deserts, and all the other features of our earth.

6.1.4 Chronological Evolution of Life on Earth

Over the past 100 years or so, scientists have developed methods of estimating the times in our earth history when certain major milestones occurred. Scientists that are involved in developing theories of evolution include paleontologists, geologists, biologists, chemists, astronomers, astrophysicists, and other disciplines. Fossil remains are the most informative in the processes of determining the history of life on earth. These remains of plants and animals that were deposited and preserved in sedimentary rocks over millions of years are now being uncovered and analyzed by scientists. Scientist can approximate the period of time in our earth history when many of these plants and animals first appeared, and when they flourished. Fossils also provide a good overview of the evolutionary process and progress of life on earth.

Scientists can determine the age of the sedimentary rock where the fossil remains exist by measuring the decay of radioactive isotopes in the rock. Radioactive isotopes are forms of chemical elements that break down to form other elements. By knowing the rate of decay of the various isotopes in the rock, scientists can determine the age of the rock, the time frame in which the fossils were deposited, and therefore when the various living organisms existed on earth. Some of the most commonly used radioactive isotopes used for dating materials are radioactive uranium and radioactive carbon, which is also known as radio carbon dating, carbon dating, or carbon-14 dating.

Scientific estimates indicate that the earth was formed 4 ½ billion years ago and the earliest life on earth appeared about 3 ½ billion years ago. Fossil records provide important evidence of the

Evolution of Consciousness

history of life. Life evolved from simple single-celled bacteria and blue-green algae into the tremendous variety of complex organisms that we have today. Excellent evidence of the evolution of life is provided in the many stratum or formations through out the world of which the Grand Canyon in Arizona is an excellent example. The location of fossil remains in the strata of sedimentary rock in the Grand Canyon shows how organisms increased in complexity through time. The lower strata (older strata) of the Grand Canyon, for instance, contain fossils of the earliest and simplest plants and animals, and the strata towards the top (younger strata) contain fossils of more recent and more complex plants and animals.

6.1.5 Chronological List of Milestones

The list below shows events in a chronological order of some of the major milestones that occurred in the history of life on earth. The order starts with the oldest and most primitive living organisms and progresses to the most recent and most advanced.

- **4.5 billion years ago** - The earth had its beginning
- **3.5 to 2 billion years ago** - *Achaean Eon*
  o Simplest single-celled bacteria and blue-green algae appeared and have flourished ever since.
- **2 billion to 700 million years ago** - *Proterozoic Eon*
  o More complex single-celled plants and animals appeared.
  o More complex multi-celled plants and animals appeared.
  o Primitive sea animals such as jellyfish and other invertebrates appeared.
- **700 million years ago** - Start of the *Phanerozoic Eon*
  o This Eon is divided into 3 eras, called *Paleozoic era, Mesozoic era, and Cenozoic era*. Life became more complex and more abundant from that time on to the present.
- **700 to 240 million years ago** - *Paleozoic Era*

- o Land plants developed and became very abundant. There were large plants in swamps. Ferns were common.
- o Many invertebrate animals developed and became abundant during this era. These included jellyfish, shelled sea animals, and colony-forming animals like corals.
- o A great variety and number of fish (vertebrates) developed. This era is called the age of fish. Lungfish and amphibians were the first air-breathing animals that also developed during this era.
- **240 to 65 million years ago** - *Mesozoic Era*
  - o Flowering plants became dominant and are still prevalent today.
  - o Snails, clams, and other shelled animals developed; many kinds of fish and amphibians also developed and flourished. Small reptiles appeared at the beginning of the era and developed to enormous dinosaurs. Dinosaurs became abundant and flourished, but became extinct at the end of the Mesozoic Era.
  - o The first warm-blooded mammals appeared towards the end of the era.
- **65 million years ago to the present** - *Cenozoic Era*
  - o Complex plants and animals that we see today came into existence during the Cenozoic Era.
  - o Ancestors of the present day horse, rhinoceros, and camel developed in the early part of the era. Dogs, cats, and antelope appeared soon after. Mammals grew larger and in greater variety around 15 million years ago.
  - o The earliest human-like creatures appeared 4 million years ago. Homo erectus appeared 1.5 million years ago. Modern Homo Sapiens first appeared about 100,000 years ago.

6.2 Darwin's Theory of Evolution of Life on Earth

Charles Robert Darwin (1809-1882), set forth his theories on evolution in his book *"On the Origin of Species by Means of Nat-*

Evolution of Consciousness

ural Selection or the Preservation of Favored Races in the Struggle for Life" which was first published in 1859. His book is more commonly known today as *"The Origin of Species"*. Darwin served as a naturalist with a British scientific expedition aboard the *H.M.S. Beagle* from 1831 to 1836, during which time he visited places throughout the world, and studied plants and animals. His study of fossils of extinct animals as well as closely related modern animals that he found in South America and the in the Galapagos Islands in the Pacific Ocean were instrumental in developing his theory of evolution. It should be noted that for the most part, prior to Darwin, the human race was unaware (except for few high level thinkers) of such a thing as evolution of life. Generally, most cultures throughout the world had their own beliefs of creation. To this day many Christians and Jews hold to the belief of Creationism as described in Bible.

6.2.1 Principles of Darwin's Theory

There are four basic ideas in Darwin's theory of evolution of living organisms as listed below. Darwin's theory specializes on the evolution of the physical bodies of plants and animals and does not particularly include the evolution of mind.

1. Evolution of life has occurred and is continuing to this day. Development of life has been from simple organisms to more complex organisms.
2. Most evolutionary changes are gradual, requiring thousands or even millions of years.
3. The primary mechanism for evolution is the process, which is known as *"natural selection"*, or *"survival of the fittest"*. Suggested here is that some members of a species are born with traits that are superior to other members of the species and have a better chance of survival. Those with the better traits survive, and also pass those traits to their offspring, providing continuous improvement of the species.
4. All life of today including the complex life forms, originated from primitive one-celled life. Darwin did not know of DNA consequently he did not know of the mechanics that con-

tributes to physical evolution. Now we know that under the right conditions DNA can mutate and causes branching of the species and deviations of life. Only the species that are well adapted to their environment survive, and those that are not well adapted will perish. Only the well suited survive to improve their lot; always from lesser to greater; always from less complex to more complex.

Darwin's theory implies that evolution of life occurs, more or less, automatically with no Divine intervention. It deals with the evolution of the physical bodies and has very little mention of mind.

6.2.2 Organization of Living Organisms

Biology is a science that is constantly evolving as the knowledge of life on earth is improved and clarified. Biologists used to classify all living things into two categories called the *Plant Kingdom* and *Animal Kingdom*. In recent years some textbooks have extended the classification to five groups of organisms since many do not fit well into the original plant or animal kingdoms.

Biologists have devised a classification system for all living organisms that is very sophisticated and detailed. This book merely lists the Five Kingdom Classification System with the general estimation of the time frame in earth history when each kingdom originated or flourished. It is interesting to note that all the living organisms listed under each kingdom originated millions of years ago and still exist today. This implies that if an organism that originated million of years ago and was successful will continue to promulgate its kind indefinitely. All life originated from the single cell, which mutated and evolved into higher plants and animals, yet the original single celled organism continues to flourish even after millions of years of existence. Many animals have come into existence, flourished and eventually went extinct, but their DNA continues to influence modern animals of their kind. For example the dinosaur flourished at one time and eventually went extinct but traces of the DNA still exist in the modern reptiles and in birds.

6.2.3 Five Living Kingdoms

1. **Kingdom Monera** – These were probably the first living organisms on earth, they are one-celled organisms that originated 3 ½ billion years ago and still flourish today. An example of a Monera is the common bacteria, some of which makes us sick and some of which are beneficial to us.
2. **Kingdom Protista** - These are multi-celled organisms that originated about 2 billion years ago and still thrive today in the form of algae and other protozoa.
3. **Kingdom Fungi** – These organisms originated about 1 ½ billion years ago and still flourish today in the form of mushrooms, yeasts, mildews, molds, athlete's foot fungus and others.
4. **Plant Kingdom** - Many primitive plants got their start 570 million years ago, and many of the more modern and complex plants started about 65 million years ago and are still abundant today. Plants include grasses, green flowering plants, ferns, and trees.
5. **Animal Kingdom** - The *Animal Kingdom* is divided into two classifications: *invertebrates* and *vertebrates*. Invertebrates are animals that do not have a bone skeleton; they usually have a hard outer covering or shell. The vertebrate animals have a bone skeleton and are more advanced than the invertebrates. Invertebrates include mollusks and anthropoids. Vertebrates include fish, amphibians, reptiles, birds, and mammals.

It should be noted that biologists, including Darwin, pay very little attention to the development of the mind in plants and animals. Generally they assume that plants do not have a mind because they do not have a brain, and they assume that the mind in animals, especially in humans, exists in the physical brain. They also assume that the developing mind is a byproduct of the developing physical bodies. These assumptions are not quite so, and we will be discussing the subject in more detail later in this chapter and in succeeding chapters.

Florencio J. Perez

6.3 Man Made Evolution

Humans have learned to emulate nature, to some extent, and have learned to produce evolution of manmade things. In many cases, humans have surpassed nature at the speed with which they have been able to make things evolve. This subject is brought up in this book to show the parallels between natural made evolution and human made evolution. The main point to make in this regard is that both nature and humans guide their respective evolution in an intelligent and purposeful manner. In both cases it may seem haphazard, but it is not; there is intelligence behind the process. There is a tendency for many to believe that evolution, by Darwin's processes of *"natural selection'* or *'survival of the fittest"*, is a mechanical clockwork process that happens by chance. The mechanism for natural evolution is in place and it runs according to Universal Law, which to some extent does run on autopilot, but it also requires guidance and supervision. Evolution is always controlled, deliberate, and intelligent.

6.3.1 Evolution of Science

The nineteenth century was the beginning of the modern rapid growth in scientific discoveries and technological innovation. The evolution of science is produced through scientific discoveries, which in turn leads to technical innovation. The scientific discoveries that have been made through history have closely tracked the highest levels of Consciousness. That is, evolution of science closely parallels the Evolution of Consciousness. As Consciousness advances it causes science and technology to advance and as science and technology advances it causes Consciousness to advance. This is another one of those feedback loops in which the growth of Consciousness accelerates technology and then technology accelerates the growth of Consciousness.

It must be noted that every scientific discovery is always based upon previous discoveries by many thinkers of the past.

6.3.2 Evolution of Tools and Technology

Evolution of tools parallels the evolution of science and technology as well as the Evolution of Consciousness. Ever since prehistoric humans first learned to make and use tools, they have been improving on them. At first these advances were arduously slow. First came the use of wooden tools, then stone tools, and thousands of year's later, metal tools. Copper was first used about 8,000 years ago. Bronze and brass, which are alloys of copper, were discovered about 3,500 years ago. The use of Iron began around 1,500 years ago. Steel, which is an alloy of iron, is a more recent development. Modern technical achievements were, for the most part, started during the nineteenth and twentieth centuries. Some of the manmade inventions that have evolved from crude devices to wonderful and dependable devices that have changed our lives are the steam engine, gasoline engine, automobile, airplane, and electronics. Other innovations include the super highways, skyscrapers, suspension bridges, ocean going vessels, nuclear submarines, hospitals, libraries, business and corporate structures, and many, many more. These are only a few of the millions of innovations that humans have created. They usually start simple and evolve slowly with time to become sophisticated and complex. The point that I am trying to emphasize by this discussion is that both natural evolution and manmade evolutions are similar in many ways, and above all they both create things in an intelligent and purposeful way. Humans tend to unknowingly copy nature when it comes to evolution. Things are not created by chance in a random order as if by coincidence. It is hoped that if the student studies manmade evolution he or she will get a better perception of the way mature works.

Unfortunately, along with all these wonderful innovations that humans have developed, there are also many negative innovations that have also been developed. Such negative things as weapon for war, war tactics, drug trafficking, criminal activities, terrorism, and many more have also been developed by humans. Such negative innovations are created because of our greed and ignorance and they eventually run their course and are superseded by better and more positive things. The Law of Cause and Effect ultimately wins over ignorance and negativity, and humanity moves forward. Nature does not develop negative things; nature only develops good things.

6.3.3 Parallels between Nature and Television

Development of the electronic version of television started in the early 1920s. Many scientists, engineers, and experimenters contributed to its development, so that there is no individual person credited with its full invention. There are several parallels that can be made between the development of television and the development of nature as follows.

- The television system is very complex and is often taken for granted; Mother Nature is much more complex and She is also taken for granted. Television is within the range of understanding of our scientists and engineers, whereas Mother Nature is still out of the range of our understanding, but huge steps are being made in the right direction.
- A person does not have to know anything about television or how it works to be able to use it. A person only has to know how to flip on the power switch and select the channel to the program desired, and then sit and watch and enjoy. By the same token a person does not have to know anything of how Mother Nature works to be able to live a normal life. Nature basically provides everything that is needed to live. The creation of a human baby is an extremely complex process; yet, a woman does not have to know how a baby fetus develops in her womb.
- A person can watch television whether the person knows or believes on the unseen parts that make the television work; unseen parts such electricity, electrons, and microwaves. By the same token a person can live comfortably on earth without knowing or believing in the unseen parts of the universe; unseen parts such as the mind, Consciousness, Soul, Spirit, or God.

- Television was developed over many years by the ingenuity of many people. The development was deliberate, worthy, intelligent and willful. It was methodical and did not happen by chance. By the same token, Nature developed life on earth in a very methodical, intelligent, and willful way and it does not happen by chance.

6.4 Nature's Guiding Intelligence (NGI)

Everything in our universe is relative and everything evolves, this include the cosmos, the solar system, the earth, all living organisms and Consciousness, and none of this evolution happens by chance. God set the Universal Laws in motion at the very beginning of time (the big bang); the main principle of Universal Law is absolute and does not change but the principles of the minor laws are relative and can change. These laws provide the operating instructions for the universe, and at first glance it may seem that the universe is off and running on autopilot with no apparent supervision. It appears that no supervision is needed, but that is not so. Behind all the evolutionary processes of the universe there is the ever-present force, which this book calls "Nature's Guiding Intelligence" (NGI).

Nature's Guiding Intelligence (NGI) is a FORCE, POWER, and ENERGY that contain DESIRE, WILL, and INTELLIGENCE to guide the evolutionary changes in an organized method. NGI is a mental will power that provides the desire and energy to introduce intelligent change in all areas of the universe. It is ever present and permeates the entire universe. It works within the boundaries of existing Universal Law, never breaking the law, but assisting and directing the change in a good way. It is always directing change in a subtle, controlled, and intelligent way. It is similar to the instinctive operating mind that operates in a subconscious manner, and so it operates under the radar of our conscious awareness and understanding. Following is a list of characteristics and attributes of NGI:

- NGI has been guiding cosmic evolution since the big bang (the beginning of time in the universe).

- NGI guides the evolution of the primitive single celled plants and animals; as well as, the evolution of the complex plants and animals; as well as, the evolution of the Human Beings.
- NGI is always at work building up and then tearing down, for the purpose of building up the material into new forms, shapes and combinations.
- NGI has three functions (1.) Creating, (2.) Preserving and (3.) Destroying. These functions apply in the Physical Domain, the Astral Domain, as well as the Spiritual Domains.
- NGI is an emanation of the Mind of God. It is a mental essence rather than physical, and it is saturated with life-energy. NGI is not a blind mechanical energy or force. It is similar to the will in humans. If a person wishes to move the arm up, he or she thinks it or wills it, and the arm moves. At first it may seem as if it was a mechanical or mindless force that moved the arm, but that is not so, it is the DESIRE and WILL of the person infused with ENERGY that moved the arm. Note the flow of proceedings; DESIRE produces WILL, and WILL produces ENERGY to accomplish the task. And, it follows that the stronger the DESIRE, the stronger is the WILL, and the more the ENERGY that is applied to accomplishing the task.
- NGI operates under the obedience of Universal Laws and limitations set by God. NGI operates throughout the universe; it appears as mechanical force, as chemical energy, as electrical energy, as gravitation etc. It is also found in all life forms including highly developed as well as primitive life forms. NGI is also allowed to operate through individual wills of individual persons, which shows up in manmade evolution of material things. *Will power* is the force that is created by *desire*.
- The force that we call *Mental Power* is the principle of the Will directed by our individual minds. The *Will Pow-*

er of the universe is always at the disposal of humans, within limits, and subject always to the laws of the universe. Those who acquire an understanding of the laws of any force may use it; but remember that any natural force may be used or misused, so be careful.
- NGI is not a reasoning force or an intellectual something. It acts as feeling, wanting, or longing; it operates as a powerful but subtle WILL in the instinctive plane of mind. NGI is similar in some ways to the Human Instinctive Operating Mind (HIOM).
- Evolution of life is constantly progressing forward toward higher and still higher forms. The urge is constantly upward and onward. It is true that some species become extinct; this happens when their work in the world has been completed. But, they are always succeeded by other species more in harmony with their environment and the needs of the times. Some races of humans also decay, but others come behind and build on the old foundations; always reaching even greater heights.
- NGI is evident with the lower single celled animals or with plants, which have no brains. It is evident in the way that insects or birds fertilize plants, or in the many ingenious ways that seeds are distributed. Seeds are distributed by the wind or water, and some plants use sweet and nutritious nectar to entice birds or insects to eat and distribute their pollen. Every species of plant has its own unique method for replication and preservation of their species. These marvelous and ingenious things do not happen by accident or by chance, they are the work of NGI.
- NGI is what helped the giraffe grow its long neck. It also helped the dog have its keen sense of smell, and the dolphin develop its sound echoing system to navigate through murky waters and to find food. It also helped the bat develop its wonderful sound echoing system to navigate in the dark. Throughout nature, NGI has helped

develop wonderful traits for the survival and evolution of the species.

6.5 The Metaphysical Account of Evolution of the Universe

Almost all ancient civilizations, tribal groups, or religious groups throughout the history of the world have had their own version of how the world came into being. It is human nature to want to know of the Supreme Being, of how the universe began, and about the origin of life in our world. The Western Religious Philosophy has been based on "Creationism" as set forth in Genesis of the "Old Testament" of the Judeo-Christian Bible. The basic story of Creationism is that God created the earth with everything in it in six days and rested on the seventh days. Modern science advocates that the origin of the universe started with a *"rapid expansion of the universe"*, better known as the big bang. Metaphysics advocates that God created the universe in His *"Infinite Mind"* and then manifested it directly from His Infinite Mind. Even though the beliefs appear to differ greatly from each other on the surface, the underlying principle in all these beliefs is similar.

6.5.1 The Metaphysical Big Bang

Metaphysics explains that in the beginning there was nothing; everything was empty space. God first created the universe in His "Infinite Mind", and then from this "Infinite Mind" emanated "God's Mind Substance" which formed the universe. This Mind Substance is made of vibrating "wavelets" as explained in Chapter 3. These wavelets are made of an infinite range of frequencies that vibrate at extremely high rate. Wavelets are imperceptible to the human senses, and it can be said that our universe is a mental universe since it is made of "God's Mind Substance".

The principal idea is that God, the Absolute, first created the universe in His Infinite Mind, during a period of contemplation. That is, God first thought out and designed the entire universe as a complete system, and when He was satisfied that the design was

Evolution of Consciousness

complete and perfect then He manifested it. The thought-form in the Infinite Mind contains the complete design of the Universal System; it included the Universal Laws, Universal Mind, energy, matter, and Life itself. It also included the complete design of the evolutionary process, from beginning to end. This "big bang" in metaphysical terms is a *"Mental Expansion"*, and Mental Expansion is the emanation of *Mind Substance*, which is made of wavelets, and it is these wavelets that make up all the components of the present universe. God's Mind Substance is what permeated the early universe, and still does today. The *"Mental Explosion"* from Gods Infinite Mind is equivalent to the scientific "big bang" theory.

Both the metaphysical community and the scientific community agree that the universe started about 13.7 billion years ago. In absolute terms, time has been but the blink of an eye, whereas from the human standpoint it has billion of years.

Evolution of the universe started immediately after the "Mental Expansion" started. A term, commonly used by both the scientific community and the metaphysical community, for the evolution of the Physical Universe is *Cosmic Evolution*. The scientific explanation of the physical part of Cosmic Evolution deals with matter and energy as discussed in a previous section of this chapter. This physical part of Cosmic Evolution is very similar as explained by both the scientific community and metaphysical communities. The similarity ceases when metaphysics explains that Universal Mind is the overriding factor that permeates the universe and is guiding every step of Cosmic Evolution. It must be noted that this Universal Mind is not the ordinary mind that we are familiar with. The mind that we are vaguely familiar with is the "Personal Mind" and it is a finite mind; Universal Mind is infinite from our present prospective.

By and large, the only part of this universe that we humans are familiar with is the physical part while the other four domains of the universe are imperceptible to us. The Physical Domain is made of matter energy, and mind. Personal Mind (finite mind) is a very small subset of Universal Mind; energy is also made of Universal Mind, and matter is made of energy substance. Physicists have made great strides in understanding matter and energy, and have mathematical equations to describe them. This is in reference to Einstein's equation $E=Mc^2$. What this says is that matter is made of highly concentrated energy. By the same token, mind is made of energy, but there is no equation yet to show the mathematical relationship between *mind* and *energy*. Mind is not physical so therefore it is out of range of our mathematics, but it is within the range of our awareness. We cannot see mind, but we know intellectually that it exists; this is similar to the fact that we cannot see air, but we know intellectually that it exists.

6.5.2 The Metaphysical "Black Hole"

The scientific theory of the *"Black Hole"* is comparable to the metaphysical equivalence called *"Cosmic Contraction"*. Evolution means creation and expansion of the universe, while *contraction* means collapse and final end of the universe. Contraction occurs at the end of the evolutionary process. This Cosmic Contraction occurs on certain parts of the galaxies at any one time and cycles through the galaxy until all is consumed while other parts are always evolving somewhere else.

Metaphysics teaches that everything in nature has a cycle. Everything has a period of activity, and a period of inactivity or rest. This principle can be seen in many things within our level of understanding. For instance, the period of day (active) and night (rest), spring-summer (active) and autumn-winter (rest), awake (active) and sleep (rest), living (activity) and so called death (rest). This can be seen in all parts of the universe. The active period of the universe is *evolution*, while the rest period is *contraction*. Each part of the cycle of "Cosmic Contraction" and "Cosmic Evolution" are extremely long periods of time. Each period lasts many billions of years.

"Cosmic Contraction" only applies to inanimate matter, and energy. It does not apply to the Human Spirit, Soul, and Consciousness. Because the Consciousness is eternal it can only evolve forward, and can never be destroyed.

6.5.3 Illustration of Systems Engineering Design

There is no way that the human Intellectual Mind can comprehend the way that God with His Infinite Mind can conceive, design, and manifest the universe. The design process He uses is infinitely complex, and there is nothing in our physical experience that can come close to that process. However, there are some crude analogies that can be used to illustration the idea. These illustrations are crude and imperfect but it is the best we can do at our level of comprehension. Two illustrations are provided. One of these illustrations involves the process that "systems engineers" use to design and build the hi-tech equipment that is so prevalent in our modern society. The second illustration involves the process that a playwright uses when writing a play for the theater.

The illustration of the systems engineer involves designing a piece of electronic equipment, but a similar scenario can be relevant when designing a building, or an airplane, or an ocean vessel, or a Space Station, etc. Usually a need for something arises first and then an idea or a mental picture is required to start the process. This is normally the seed thought that starts the process. From the idea the "systems engineer" develops a requirements document that defines what the gadget is expected to do. The next step is a planning stage, which involves developing a conceptual design. The conceptual design defines such things as the most feasible approach to take and the latest technology to use. It also defines staffing and experience mix required. Also included are schedule and budget estimates for each phase of the project, which includes design, development, test, and implementation.

The next step after the conceptual design is the detail design. This is where the engineer has to focus his thoughts to the smallest detail. Every last detail has to be thought out completely; if even the slightest detail is not thought out completely, the equipment is destined to fail. Nothing can be left to chance. These thoughts are all recorded on paper in the form of sketches, drawings, mathematical equations, specification documentation and much more. The result of the detail design is a complete recorded definition of every aspect of the equipment to be built. If any thing is overlooked, forgotten, or in error; the equipment will not work; in engineering design there is no room for sloppy work. A good detail design lays the foundation for the rest of the project. The steps after the detail design are: procurement of material, fabrication, assembly, testing, installation, and system integration testing. It takes the work of many highly trained and motivated people in a highly organized environment to accomplish a good system design. All these, of course, have to be within the framework of the laws of physics. The final product is a physical manifestation of the original need. This physical product was originally created in the mind of the systems engineer and his engineering staff. It is important to note that all physical manifestations are always first created in the mind of someone, and then the desire and will are required to make it happen. Nothing ever happens by chance. This scenario of the Systems Engineer is a crude presumption to attempt to explain the unexplainable. It is a crude conjecture of the infinite. God's methods of manifesting a universe are infinitely more complex than what this scenario can convey.

6.5.4 Illustration of a Playwright

Another crude illustration that might help to give a vague mental semblance of how God designs a universe is by examining the work of a playwright. A playwright starts with an idea that is usually based on some past experience. This idea forms the thought-form of the story that the playwright forms in his or her mind. These thoughts are then translated into words that are documented on paper. The story as documented contains the script, setting, detail character descriptions, plot, etc. A complete theater production is very complicated; it involves a director, performers, costume designers, musicians, choreographers, and financial and business planners. When the play is performed on stage, it may seem real to some since it is real people, with real human characteristics, emotions and goals that are performing. But, in actuality it is fiction (comparable to an illusion), which was created in the mind of the playwright.

We can speculate that God uses methods that are similar to the illustrations mentioned above, but keep in mind that His methods are infinitely more complex than these illustrations can portray.

6.6 Evolution of the Instinctive Operating Mind

The "Infinite Mind of God" is an absolute mind and is not of this universe, but it is the mind that emanates everything in the universe. Another way of saying this is that the universe is of the Infinite Mind but it is not the Infinite Mind. The "Infinite Mind of God" is absolute and unknowable. It is unknowable to us since we have a limited and finite mind. Universal Mind is an emanation from Infinite Mind, but it is not an absolute mind. Universal Mind is a relative mind that originated with the initial "Cosmic Expansion" and is the basis for all the other minds of the universe. It is the master organizer, builder, and operator of the universe. All the billions of minds that make up the universe are subsets of the Universal Mind. All energy, matter and finite mind are made of Mind Substance from Universal Mind, and all Mind Substance is made of wavelets.

As mentioned, there are billions of minds in the universe but the focuses is on those minds that are most directly relevant to the theme of the book. The minds that are most important to us at this point are the instinctive operating minds. All instinctive operating minds function only in the Physical and Astral Domains and do not have a function in the Spiritual Domains since there is nothing physical there. For illustration purposes the instinctive operating mind can be crudely compared to the operating system (OS) of the modern computer. They both operate the system below the radar of conscious awareness.

Instinctive minds are subconscious minds that exist throughout the Physical and Astral Domains. They exist in very low-level versions to very elevated versions. They exist in inanimate matter with no Consciousness as well as in living organisms that have Consciousness. They progress from the most elemental form of instinctive mind to the very advanced. The instinctive minds are listed below in ascending hierarchical order:

- Mineral Instinctive Operating Mind (MIOM)
- Terrestrial Instinctive Operating Mind (TIOM)
- Galactic Instinctive Operating Mind (GIOM)
- Plant Instinctive Operating Mind (PIOM)
- Animal Instinctive Operating Mind (AIOM)
- Human Instinctive Operating Mind (HIOM)
- Nature's Guiding Intelligence (NGI)

In this list the MIOM is the most elemental and NGI is the most advanced. The MIOM, TIOM, and GIOM operate the functions of inanimate materials, which are defined as non-living material with no Consciousness. PIOM, AIOM, and HIOM operates the physical body functions of living organisms, all of which have Consciousness in varying degrees of advancement. All instinctive minds operate or manage the functions of entities in their particular area of expertise, but they can also be mixed and matched according to the function necessary. For instance, MIOM is considered to work in conjunction with PIOM, where MIOM handles the atomic functions of molecules of the cell, and the PIOM handles the living functions of the cell. There are a multitude of instinctive minds that operate at the level of the Spiritual Domains but they are presently unknown to us and therefore not covered here.

6.6.1 Relationship of Universal Mind and Universal Law

Universal Mind as well as all its subservient minds is relative and they continually change, whereas Universal Law is absolute and never changes. Universal Law has been in operation since the beginning of time without change and has been using mind to carry out its directives. In essence there is only one prime Universal Law that permeates the universe, but we break it up into several Universal Laws and numerous sub-laws for the purpose of ease of understanding and explanation. What we call Universal Laws are completely interdependent, interrelated, interactive, and inseparable because they are all but one law. Only five entities are identifiable

Evolution of Consciousness

as absolute; they are God, God's Infinite Mind, Spirit, Soul, and Universal Law, all else in the universe is relative.

6.6.2 Evolution of MIOM

The most elemental or primordial instinctive mind is the Mineral Instinctive Operating Mind (MIOM). This is the mind that automatically runs the functions of inanimate matter at the subatomic, atomic and molecular level. There is no Consciousness associated with inanimate matter. MIOM was the first mind in operation after the "big bang". It supervises and controls the functions of subatomic particles such as the quarks, photons, bosons, and all the rest. It defines the frequency vibrations as well as the properties and characteristics of the super-strings. It also controls the atom at the nucleus and orbital levels of the electrons. The more advanced MIOM supervises and controls the functions of molecules from the fairly simple atomic combinations, such as H_2O, to the complex combination such as hydrocarbons, proteins, hormones, and DNA itself. The MIOM evolved from controlling the simplest atom such as the hydrogen atom to the most complex molecule such as DNA. The fundamental laws of physics and chemistry at the microrealm are expressed by the criterion of MIOM. It is amazing that MIOM started operations in primordial times, at the beginning of the universe, and are still in operation today. MIOM operates all matter at the atomic level in both the Physical Domain as well as in the Astral Domain; it operates the atoms and molecules in all living cells of living organisms to include plants, animals, and humans.

6.6.3 Evolution of TIOM

The Terrestrial Instinctive Operating Mind (TIOM) operates all the functions of our earth with the earth being one of the most advanced planets in the galaxy. The TIOM operates all the weather dynamics of the planet as well as the flow of ocean current throughout the world. It also controls the dynamics of the inside of the earth such as the movement of the liquefied core and magma flows that created volcanoes, plate tectonics, earthquakes, and all the rest.

TIOM maintains the exquisite balance and correct proportionality of atmospheric oxygen, nitrogen, carbon dioxide, water vapor, and all the rest of the gases in the air. TIOM controls the earth's rotation on its axis, the tilt of the axis that generated the seasons, its orbit around the sun as well as the moon's influences on earth. The TIOM controls and operates all the functions of the earth in an instinctive and semiautomatic fashion.

Every moon, planet and star in the universe has an instinctive operating mind similar to the earth's TIOM, but the earth's TIOM ranks among the most advanced because our earth is a very advanced planet. Planets are generally ranked hierarchically in accordance to the level of activity of the planet. For instance our moon being very inert has a relatively low version of TIOM, and our Earth having flourishing life has a very advanced version of TIOM.

6.6.4 Evolution of GIOM

The Galactic Instinctive Operating Mind (GIOM) controls the overall organization of the planets, moons, stars, black holes, and other objects within the Milky Way Galaxy. It balances the distribution of gravity, matter, and energy of the galaxy in both the Physical and Astral Domains. MIOM, TIOM, and GIOM organize and operate material that is non-living (inanimate material with no Consciousness). All galaxies in the universe have their own version of GIOM to operate their basic functions in an instinctive and semiautomatic mode. These instinctive operating minds work in harmony with each other. They work in a semi independent format to a certain extent but always under the supervision of Universal Mind and NGI.

6.6.5 Evolution of PIOM in the Plant Kingdom

The Plant Instinctive Operating Mind (PIOM) was the first instinctive mind to operate the functions of the very first living organisms on earth. It was the first mind to sustain life on earth. All living things have Consciousness, so Consciousness started with these primordial organisms. The earliest and simplest PIOM started

Evolution of Consciousness

the operation of the first and simplest single celled organisms, and after that it progresses gradually in degrees up the scale of evolution. With each degree of development PIOM got more complex and sophisticated. The scale of PIOM development ranges from the very primitive plants to the very advanced plants. The most primitive forms of life are the single-celled organisms, protozoa, algae, yeast, fungus, etc. The most advanced plant life of today includes the grasses, trees and flowering plants. Plants do not have a brain for centralized command and control, consequently PIOM is distributed throughout the plant.

PIOM is what keeps the plant alive by operating and managing all its vital functions at every level of plant life, that is, it operates at the cell level as well as at the overall plant level. At the cell level it manages and operates each individual cell of the plant, it operates the cell's food assimilation and elimination as well as the cell's energy generation and usage. It handles the cell's reproduction by division, plus all the thousands of other things that have to be done to keep the cells alive and healthy. All the billions of living cells of the plant are managed and operated simultaneously and in coordination with each other. You might say that the plant invented the cell operating procedures that are used by all living things of today including the cells in humans. MIOM works in conjunction with PIOM. MIOM handles its regularly assigned function of operating the atoms and molecules, and PIOM handles all other function of the plant above the molecular operations. MIOM controls the DNA from the molecular standpoint, and PIOM controls the DNA from the cell's reproductive standpoint. All minds coordinate their processes and functions so that they are all in sync, in agreement, and in harmony with each other's efforts.

At the plant macro level, PIOM manages the plant food gathering system; where it gathers water and minerals (nutrients) from the ground through its root system. It also distributes the nutrients throughout the plant, and manages the process of photosynthesis, which is the process that produces energy and food with the aid of the sun. PIOM also supervises the seed's reproduction and distribution process.

The PIOM varies widely in function from plant species to plant species depending of the type of plant it serves. It varies in degrees of sophistication from the very primitive plants to the modern flowering plants. Each plant has its own version of PIOM; some versions are closely related whereas others are very different. Yet, every plant has its own individual PIOM that best suits its needs for life on earth. PIOM evolves forward under the guidance of NGI, and as a result of PIOM advancing forward, it causes the physical plant to also advance and evolve. The physical plant evolves forward due to the prompt of the PIOM. As the PIOM evolves it generates new requirements for its physical counterpart and with the help of NGI induces the physical plant to change to meet the new requirements. There is a common misconception that the evolution of the physical organism induces the mental to change, but that is not so; it is the mental that stimulates the physical to change.

6.6.6 Evolution of AIOM in the Animal Kingdom

The next higher level of life on earth after the Plant Kingdom is the Animal Kingdom. Animals have an instinctive operating mind that is much more advanced than even the highest PIOM of the most advanced plants. The instinctive mind of the animals is called Animal Instinctive Operating Mind (AIOM). AIOM is composed of MIOM and PIOM plus the new additions of AIOM. The MIOM handles its usual function of operating atoms and molecules throughout the animal's body, the PIOM handles its usual function of operating the living cells, and the AIOM handles the new functions of the animal. When Mother Nature invents a useful and dependable process that works well in a lower form of life, she tends to use the same or modified process again and again on higher forms of life if applicable. In this case she uses the same basic cell management process of the plant, in the animal. She also uses a similar food and mineral distribution process of the plant, in the animal.

The animal and its AIOM made huge advances over the plant and its PIOM. A few of the advances of the AIOM are as follows:

Evolution of Consciousness

- Development of a brain and nervous system. The brain is physical and is used as a centralized command and control system in animals. The AIOM is non-physical and directs most the functions of physical body through the brain. The AIOM issues commands to the brain, and the brain converts those commands to electrochemical (physical) signals, which are distributed to the rest of the animal's body through the central nervous system.
- Developed mobility. The AIOM directs all motion of the animal in an automatic or instinctive manner.
- The AIOM directs the respiratory system. (Gills for fish and lungs for land animals).
- The AIOM directs the circulatory system, immune system, digestive system, sensory system, elimination system, reproductive system, and others.
- The AIOM developed two very strong instincts, (1) the will to live (survive), and (2) the urge to reproduce. These critical instincts were actually started in the Plant Kingdom and only become more dominant in the animals.
- The AIOM also developed the new animalistic emotions, feelings, and passion. The emotional system was created for the survival of the animal. Such emotions as fear, hate, anger, cruelty, rage, and other negative forms of emotions were necessary for the animal's early survival. A rudimentary semblance of some of the positive emotions such as affection, joy, kindness, and caring for family members was also developed in the higher mammals. The importance of the animalistic emotions is that they became a permanent fixture of the AIOM and humans inherited those negative traits from the AIOM.

The AIOM starts very simplistic in primitive animals and advances through evolution to high level of sophistication in the higher mammals. The AIOM varies widely in function from one animal species to the next. It can be said that each species has a different version of the master AIOM copy. The more advanced animals have a more sophisticated and advanced AIOM than the lower animals. Each individual animal has its own individual AIOM that is particularly suited for its level of progress. AIOM advances with the prodding and guidance of NGI. AIOM along with NGI than prods the physical to according to its new requirements. This is all done in conjunction with Darwin's principle of "survival of the fittest".

6.6.7 Evolution of Human Instinctive Operating Mind (HIOM)

The HIOM is the most advanced of all instinctive operating minds. It is the summation of MIOM, PIOM, AIOM and the new parts of HIOM. The MIOM does its usual function of operating at the atomic level and the PIOM does its usual functions of operating the living cell. The AIOM does its usual function of operating all the autonomic functions such as the respiratory system, the circulatory system, the immune system, the nervous system, and so forth. In the beginning of human evolution the HIOM is not much different from the Animal's AIOM. The big advance of the HIOM over the AIOM is that HIOM has a new *"Trainable Mind"* that the AIOM does not have. The new Trainable Mind is the one that allows the human to learn to use tools, to drive a car, to play the piano, to ride the bicycle, and to learn the thousands of other things that humans do and animals cannot do. The Trainable Mind is also responsible for the human speech.

It is the Trainable Mind along with the *"large trainable human brain"* that makes it possible for humans to talk and to do so many physical activities. The Trainable Mind operates under the direction of the Human Intellectual Mind until it learns to do the activity well. After the Trainable Mind learns to do the specified action well, it no longer needs the Intellectual Mind to give it detailed direction anymore. The Trainable Mind learns and gains proficiency with repetition. After the Trainable Mind learns the activity the Intellectual Mind is relived of the drudgery of repetitive activity. The Trainable Mind is good at doing repetitive activities whereas the Intellectual Mind is not. The Trainable Mind directs the physical brain to do the activity and the brain learns to do an activity through repetition much as the Trainable Mind learns through repetition. The brain in turn instructs the muscles to do the activity and through repetition the muscles also become proficient. The end result is that through repetition the Trainable Mind, the brain, and the muscles learn to coordinate the activity and become proficient without the need of the Intellectual Mind. The HIOM, the Trainable Mind and the brain are not thinkers they are doers of physical activities only. Their only function is to produce physical actions in the physical body. The HIOM, the Trainable Mind, and the brain have no function in the Spiritual Domains since there are no physical bodies there.

Another very important part of the HIOM is that it contains the animalistic emotions, feelings and passions that were acquired by the animals during their early evolutionary period. These negative emotions were inherited from the AIOM and are a permanent part of the HIOM. These emotions were needed for the animal's early survival and were also needed by the early humans for their survival. As we developed, we no longer need these negative emotions, and they have become a burden on us. It is now our job to become aware of those negative emotions and through "force of will" suppress their effect. The negative emotions will be conquered when we become aware of them and then convert them to positive emotions; in other words, we will win when fear is converted to courage, anger is converted to calmness, greed is converted to restraint, cruelty is converted to kindness, et cetera. Refer back to Chapter 2 for more information on the HIOM.

6.7 Evolution of Consciousness

God did not create the universe to have moons, planets, and suns; he created the universe to provide a platform for Consciousness to evolve in. Consciousness is the most important entity in the universe. Life requires Consciousness to exist; Consciousness exists in all living organisms from the most primitive to the most advanced human. Non-living materials such as rocks, water, air, moons, planets, suns, galaxies and such do not have Consciousness. Non-living materials have mind but do not have Consciousness. The most primitive Consciousness exists in the most primitive life form. They were the first organisms to live on earth in its most primordial times. From that time on, Evolution of Consciousness has proceeded unabated, from the most primitive to the most advanced human of today. Consciousness is relative thought out the evolutionary process.

6.7.1 Parallel Evolution of Physical Body and Consciousness

The evolution of living things on earth encompasses the parallel evolution of instinctive mind, physical bodies and the Consciousness. This is so, for all life on earth from the simplest single cell plants and animals to the most complex Humans. The main objective of Mother Nature is the development of the Consciousness. The evolutionary development of the instinctive mind and physical bodies is of a secondary aim of nature. Physical bodies and minds are provided as tools for the Consciousnesses. Physical bodies and instinctive mind evolve from lower to higher forms, but the driving force for their evolution is the requirement of the Consciousness in conjunction of NGI. As the Consciousness develops forward, it requires a comparable advancement in the mind and physical body. So it is the combination of the needs of the Consciousness, along with the help of NGI, which drives the advancement of the development of the instinctive mind and physical body.

In the cycle of life on earth every physical body dies off after completing a lifetime. This applies to all life forms from the simplest to the most complex. In the physical assemblage the information of the species is carried from generation to generation in the genetic structure of DNA. Evolutionary changes in the physical occur when the DNA mutates ever so slightly. A mutation of the DNA usually means that a few molecules rearrange themselves within the long helix molecule in the reproductive cell of the plant or animal. This mutation does not happen by chance; it is always an intelligent change. It is prompted by the desire of the Consciousness to advance the tools where it will continue to grow. NGI is always present to facilitate a change in evolutionary growth. It is an intelligent mental

Evolution of Consciousness

force that contains *energy* and *will power* to make changes in the life forms. These changes are always in the positive direction. Even the unsuccessful iterations of plant and animal species are important to the overall plan of evolution.

6.7.2 Evolution of Plant Group Consciousness

Plants have Consciousness, but it is in the form of a "Group Consciousness". What this means is that all plants of a particular species belong to one Group Consciousness. For example all oak trees belong to one Group Consciousness, all tomato plants belong to one Group Consciousness, all kelp-seaweed plants belong to one Group Consciousness and so on. The Group Consciousness of plants began at a very primitive stage but gradually developed in sophistication as it developed up the ladder of evolution. The Group Consciousness of each species of plants actually gains experience and knowledge with the life of each plant that lives a lifetime under its jurisdiction. It takes the experiences of millions upon millions of plant lives, of a particular species, to accumulate sufficient knowledge, for the Group Consciousness of that species to advance a notch. Awareness level of primitive plants is nonexistent or zero, but no matter how insignificant, it represents a start for Consciousness.

6.7.3 Animal Group Consciousness

Animals also have Consciousness, and like plants they have "Group Consciousnesses". Animal Group Consciousness is much more advanced than Plant Group Consciousness. Each animal species has its own Group Consciousness. For example, all rainbow trout belong to their own Group Consciousness, all rattlesnakes belong to their own Group Consciousness, all peregrine falcons belong to their own Group Consciousness, all rabbits belong to their own Group Consciousness, and all chimpanzees belong to their own Group Consciousness, and so on. The Group Consciousness of animals began at a very primitive stage and developed one step at a time over millions of years to the level of sophistication of today's primates. The lifetime of each animal on earth contributes to form the overall knowledge base of the Group Consciousness of the species. This gain in knowledge is cumulative over millions of years, and the overall result is the advancement of the Group Consciousness. A Group Consciousness evolves within a species and eventually it reaches the highest point of development within that species; when that happens, a chunk of it breaks away from the main and moves to the next higher animal species on the scale of evolution. The awareness level of the lower animals is mere sensation whereas the awareness level of the higher mammals goes up to the level of awareness of the physical. That is, the higher mammals are aware of their physical bodies from the standpoint that they feel pain, cold, heat, hunger, discomfort of insect bites, etc.

6.7.4 Analogy of Dewdrops to Clarify Group Consciousness

An analogy used to help illustrate and clarify this idea of Group Consciousness, is as follows: Suppose there is a large lake that produces millions of tiny dewdrops as a result of evaporation and condensation. The dewdrops are then distributed over a large area around the lake; some are deposited on blades of grass, some on leaves of plant, some on rocks, et cetera. Wherever they end up, each dewdrop picks up tiny bits of minerals such as salts and dirt that dissolve and goes into solution in each dewdrop. In due time all these dewdrops eventually drain back to the lake carrying back all the foreign material in solution. At that time the foreign material becomes the property of the lake. The next day or so, more dewdrops are distributed and eventually they also return to the lake carrying more foreign material. Each generation of dewdrops distributed carries back additional material, which becomes the property of the lake, and the process continues on and on. Gradually the composition of the material in the lake changes with each successive generation of dewdrops.

In this analogy (illustration) the lake represents the "Group Consciousness", and the dewdrops with minerals in solution represent the experiences gained by the plants or animals as they lived their lives on earth. This story represents the development of Consciousness in primitive plants and animals, and then progresses to higher and higher plants or animals. Plants and animals are not permanent independent Consciousnesses, but are families of "Group Consciousness". From this families of "Group Consciousnesses" gradually break off minor groups, which form other species, or subspecies, and continue on and on.

This is so until the Human Beings steps into the picture. The Human Being does not belong to a "Group Consciousness". Each Human Being is its own individual and independent Consciousness. When the higher forms of animals reach the highest point of their evolutionary process; they reach the plane of the Human Being. At that point the "Group Consciousness" of the animal breaks up into permanent individual Consciousnesses, which signifies the beginning of the Human Being. Also, at this point, true reincarnation begins. What this means is that now each Human Consciousness becomes permanent, eternal, individual, and independent entity that is set on the path of Evolution of Consciousness. And that is what this book is about.

6.8 Evolution of Human Consciousness

Human Consciousness represents a quantum leap over the animal Group Consciousness. Humans do not have Group Consciousness; Humans have one Consciousness per individual. Human Consciousness starts when the Animals Group Consciousness reaches its maximum level of its evolution. At that time a certain section with the highest attributes of the Group Consciousness breaks out and becomes a Human Consciousness. The new Human Consciousness enters a human baby of a primitive human tribe and this represents the first birth as a human. At that time the new human also acquires a complex Personality, the Intellectual Mind, the Intuitive Mind, the Soul and the Spirit. From those primitive beginnings the human is on its way to true human development. Chapter 8 contains more details on the stages of Evolution of Human Consciousness.

6.8.1 Launching the First Humans

There is a lot of speculation and controversy about the origin of the first human being on planet earth. Anthropologists and Archeologists have been doing a lot of work along these lines and have come up with many good theories. They study ancient cultures of humankind through examination of material remains such as ancient graves, tools, and artifacts. Paleoanthroplolgists study skeletal remains such as fossilized human bones and skulls to try to piece together the early beginnings. These theories are generally on the right track, but there is always widespread disagreement among scientists. Scientists do their work from an objective and observational standpoint; they tend to frown on using subjectivity on their methods. The scientific method of investigation is commendable and has made many important discoveries that contribute tremendously to the knowledge base, which has freed modern humankind from much superstition and myth. On the other hand, pure objectivity tends to handicap scientists and causes them to miss some important points. Scientists normally do not take into account mental evolution, Spiritual evolution, or motives of evolution. These aspects should be included as parts of the scientific investigations in order to further understand the problem.

The animal's physical body and instinctive mind is essentially the same as the human. But once you get past the body and instinctive mind, the human represents a quantum and profound leap over the animal. The general theory that the human physical form was derived from evolutionary branches of the great ape is correct. I believe that one of the evidences for the theory on the evolution of the great ape is made by the find by Donald C. Johnson in East Africa in 1979. He found remains of the oldest and most complete hominid (human ancestor) skeleton ever recovered and nicknamed it Lucy. Lucy was a female skeleton that measured about four feet nine inches tall, walked upright, and had a small apelike head. Lucy has been dated at around 3.2 million years and had a brain size of about a quarter to a third of our modern brain size. Judging from the size of Lucy's brain she was not yet a human at 3.2 million years ago but she already had many of the features required by the human.

Evolution of Consciousness

She was indeed a forerunner on the way to becoming a human, but even though she walked upright she only had an advanced version of AIOM. She did not yet have the HIOM, or intellectual capacity of the human.

I propose that it was a relatively short time after Lucy that the true human developed. The true human is a huge advancement over the animal and has to have at least nine new constituent parts that even the most advanced animals such as the great apes do not have. These parts, in the order of their development, are as follows:

1. Bipedal upright walking to free up the hands
2. Individual Spirit and Soul
3. Individual Consciousness
4. Intellectual Mind
5. HIOM with Trainable Mind
6. Large brain
7. Physical agility for using tools
8. Speech
9. Complex Personality

Upright walking was accomplished in preparation to the acquisition of the new human parts because upright walking is an important prerequisite for further development. The importance of upright walking is that it frees the hands to do more things which in turn facilitate greater capacity to gain new experiences. The logic of this principle is that Mother Nature requires that the human gain experiences in order to develop and evolve forward. All the new parts are designed to do just that; they are all interconnected to provide a feedback loop where each constituent part requires the development of the next constituent part. Following is a brief descriptive reason for the need of each constituent part.

6.8.2 Explaining the New Human Parts and Functions

Walking upright is necessary to free up the hands for doing things because doing things gains experiences that are needed for evolutionary progress to take place. Walking upright is a bit of a handicap for humans because locomotion on two legs is not as fast as locomotion on four legs, but humans more than make up for that deficiency with their superior intellect. Walking upright frees the hands to do things such as making and using tools, carrying materials for shelter, making and carrying weapons for hunting or for defense, gathering food and much more. In other words the higher intellect and freeing of the hands to do things, opens up a brand new world of experiences. Walking upright was not the cause for becoming human, it was the prelude to becoming human and it was anticipated, willed, and promulgated by NGI. Animals do not have two free hands because there Consciousness is not advanced enough to be able to benefit by the use of two free hands. At their level of development their Consciousness is till growing without the need for two hands. Mother Nature provides the needed equipment only when it truly becomes a requirement for continued growth.

In order for the human to become a *Real Self* it had to acquire the independent Human Spirit, Soul, and Consciousness. This was a major accomplishment of Mother Nature and it was not an accident. It was planned, premeditated, and constituted a major milestone. Remember that Mother Nature can do anything she wants to, whenever she wants to. Her designs are always geared to the single purpose of Evolution of Consciousness, which is always her number one priority.

The next requirement after the Real Self is installed in each human is the creation and installation of the Intellectual Mind so that the Consciousness can have its instrument for thinking, analysis, and decision-making. The Intellectual Mind is not a part of the Real Self but is used by the Consciousness as a tool to think with. The Intellectual Mind is always attached to the Consciousness and evolves at the same pace as the Consciousness. It is always at the same level as the Consciousness. The Intellectual Mind is a thinking mind as opposed to a doing mind. Because it is a thinking mind it is not good at doing repetitive things; repetitive things are drudgery to the Intellectual Mind. The Trainable Mind is then created to learn to do repetitive activities under the supervision of the Intellectual Mind and thus alleviating the Intellectual Mind of drudgery. The Trainable Mind is not a thinking mind, but it is good at doing repetitive things;

it learns through repetition. The Trainable Mind becomes a part of the HIOM, which does things in a more or less automatic mode. The Trainable Mind is a wonderful mind, but it does pose a problem in that it learns all that it is told to learn including the good as well as the bad. It does not know how to differentiate between good and bad, right and wrong. Learning good habits is not a problem, but learning bad habits is a real problem because bad habits are hard to reverse.

The creation of the Trainable Mind in turn requires the creation of a larger brain to do all the many more physical activities that are required. All minds are non-physical but the brain is a physical device, which would indicate a situation of non-compatibility, but that is not so. The compatibility between physical and non-physical is accomplished by an interface mechanism that converts and relays messages between the non-physical HIOM and the physical brain in both directions. The larger brain is not a thinking instrument; it is strictly used for doing physical activities. The brain is used as the central command and control hub to direct the muscular activities of the physical body. The instinctive mind does not necessarily require a large brain; it is the huge number on new things that we do as a result of the learning capacity of the Trainable Mind that requires the large brain. For example it is as a result of the Trainable Mind that we can learn to ride the bicycle, learn to drive the car, learn to play the guitar, learn to use tools, learn to write, and learn to do thousands of other things. It is also because of the Trainable Mind along with large brain that we can learn to talk. Speech requires a large brain capacity because of the physical complexity of speech.

Once the hands are free as a result of the upright posture, and the Intellectual Mind, the Trainable Mind, and the large brain are functional, the human is ready to start doing all the many things that humans do. The new human can now learn to make tools and weapons for hunting and fishing or defend its family and tribe against intruders or wild animals. Mother Nature's ulterior motive is to give the human the ability to do many things because doing things expands the human's ability to gain experiences that are necessary for its further development and evolution. Mother Nature's ulterior motives are always designed for the purpose of expanding the capacity of the Consciousness to learn and therefore unfold.

6.8.3 Speech

Part of the expanded capability to do more physical activity is the ability to talk. With the Trainable Mind and larger brain the human now has the capacity to learn to talk. Speech is an extremely complex physical process, which requires the coordinated activity of many muscles and body systems. For instance, in order to say a word, any work, it takes the simultaneous and coordinated movement of the tongue, cheeks, lips, jaw, nasal cavity, voice box, lung and breathing synchronization and many more. The ability to string many words together into sentences is even more complex, and that is why speech necessitates a large brain. Animals are confined to simple grunts, or squeals, or squawks because they do not have the brain capacity to do much more but that is all that they need for their level of development. The combination of the Consciousness, Intellectual Mind, HIOM and large brain also gives the human the ability to acquire a complex personality, which is also necessary for the development of humankind.

6.8.4 The First Human

The development of the true and complete human took place about 2.5 million years ago, and it took a relatively short period of time to accomplish the feat. It probably took in the order of 50 to 100 generations to achieve the transition. The reason it was so quick to take affect is that Mother Nature had already done all her homework and was ready to put the plan into action. Mother Nature can do things that are unfathomable to us but which are normal and commonplace to her. Once the human race was complete and functioning well, it started its evolutionary journey forward, never to look back. Also note that the human physical body did not change much after its original beginning, it may have changed some in size, and strength, and speed, and in skin pigment coloration but the essential body form has not changed much. What has changed tremendously from the very beginning is the Consciousness. The Consciousness has been evolving from its primitive beginning to the advanced stage that we are at now and we still have a long way to go. Chapter 8 has the details of the various stages of development that are still ahead for us in the future.

In both plants and animals, the Group Consciousness advanced by moving from a lower species to the next higher species, but It should be noted, that in humans there is only one species; only one body type. When the Consciousness reaches the level of the human, one body type fits all human evolution. The one and only human body type has all the necessary capacity to accommodate the evolution of the Human Consciousness on earth. The human body has reached the zenith of its physical evolution, and the Human Consciousness moves up from level to level using that one-and-only and incomparable body type.

6.8.5 Super-men evolving into Super-Gods

There is a story from the Eastern Masters that goes like this, "Men are evolving into super-men; super-men are evolving into gods; gods are evolving into super-gods; and super-gods into something even higher; there is an infinite ladder of being; the Infinite Mind of God is in all, and all is in the Infinite Mind of God." These gods and super-gods are not to be confused with God the Absolute. What the masters mean by gods and super-gods, in this case, is that there are Consciousnesses that are so highly advanced and so far above our human level of comprehension, that if we were to encounter them consciously we would think they were gods. In reality they are not gods; they are Consciousnesses just like we are, except that they are far more advanced than we. There are not enough outlandish superlatives to describe their goodness, power, and greatness. The wonderful part about it is that we are on the path to eventually become just like them.

Chapter 7

The Physical and Astral Domains Of the Universe

The universe is actually one Universal System made of many components, which work in perfect unison and harmony. Every component is inextricably linked with every other component; there are no extra components and there are no missing components; the system is perfect. God created the universe in His Infinite Mind, and then He manifested it into being from that Infinite Mind; therefore the universe is a mental universe.

This chapter will focus on explaining the characteristics of the Physical and Astral Domains, and the next chapter will discuss the characteristics of the three Spiritual Domains. The Physical Domain is the only domain that we know well and consequently does not require an extensive explanation. However, we will discuss it in sufficient detail to provide a foundation for discussion of the more advanced domains that follow. Every domain has its place and function in the universe and even though their functions are different from one another, they also have many similarities. The discussion of the Physical Domain will set the foundation for forthcoming discussion of the higher domains. One of the important characteristics of the universe is that the universe is hierarchical, and by far the most important entity in the universal hierarchy is the Consciousness. Much of the discussion in this and the next two chapters will deal with the hierarchy of the domains and Consciousness.

Domains – The universe is made of five domains. Spiritual Domain 3 is the most advanced domain in the universal hierarchy, and the Physical Domain is the least advanced. What our society knows as the "physical universe" is called the "Physical Domain" in this book. All five domains are occupied by Consciousnesses, and Consciousness is what gives the universe its reason for existence. The domains in descending order are as follows:

- Spiritual Domain 3 (SD3) – This is the most advanced domain in the universe.
- Spiritual Domain 2 (SD2)
- Spiritual Domain 1 (SD1)
- Astral Domain
- Physical Domain – This is the least advanced domain in the universe.

Consciousness – The main purpose for the existence of the universe is to provide a platform for Consciousnesses to develop. Consciousnesses are segmented into stages that represent broad categories of advancement. Consciousnesses range from the ultra primitive to the ultra advanced. The following is a list of the six stages of Consciousness in descending order.

- Spiritual Consciousness Stage – This is the highest stage of Consciousness.
- Cosmic Consciousness
- Soul Consciousness Stage
- Human Consciousness Stage – We are at the upper levels of this stage now.
- Animal Consciousness Stage
- Plant Consciousness Stage – This is the lowest stage of Consciousness.

Planes and Sub-plane: All domains are subdivided into planes and sub-planes that represent a hierarchical order. The least advanced planes are always shown at the bottom and the most advanced planes are shown at the top of the diagram. Planes and sub-planes are not actual locations or places; they are states of attainment. We tend to depict planes in terms of up and down for the sake of explanation and illustration but in actuality there is no such thing as up and down in the Spiritual World because they are states of being rather than physical locations. Each domain is composed of millions of planes and each plane is composed millions of sub-planes. Consciousnesses naturally migrate toward the plane that is the most suitable to its needs according to its level of development. Consequently each plane contains Consciousnesses of more or less the same level of development.

7.1 The Physical Domain

The Physical Domain is the domain that we know best and it is the lowest domain in the hierarchy of the universe. Like all the other domains, it is divided into many planes and sub-planes. Refer to Diagram 3. The list of planes in the diagram is arranged in descending order (from higher to lower). The least advanced or least significant planes are placed at the bottom of the diagram, while the most significant or most advanced planes are placed at the top of the diagram. The diagram is for illustration purposes only and not to be construed as conforming to some accurate universal scale. Other scientific groups such as biologists, chemists, geologists, and astronomers would perhaps divide the physical planes differently. The explanation that follows on the planes of the Physical Domain are intended to set the tone for explaining the planes of the more advanced domains, which are not as well known.

7.1.1 Planes of the Inanimate Physical Domain

The planes of the inanimate universe do not have Consciousness associated with them and are by definition not alive and therefore they are the lowest planes in the hierarchy of the universe. No

Consciousness, of course also means zero level of awareness. The inanimate universe does however have instinctive mind that operates its functions according to the laws of physics. Each category of planes described below represents perhaps millions of planes.

Planes of Subatomic Particles, Atoms, and Molecular – These planes and its sub-planes pertain to the *micro* world of the Physical Domain. This micro world includes the subatomic particles, atoms of the elements, and molecules. These molecules include the simplest molecules as well as the compound molecules such as DNA. The MIOM operates at this level and is considered to be the lowest instinctive mind of the universe. Yet, it contains all the functioning knowledge and operations of atomic physics, nuclear physics, particle physics, and molecular physics and chemistry. The latest methods that physicists are using to explain the micro-world of physical matter and energy are Quantum Mechanics and the Super-String Theory.

Planes of Celestial Bodies – These planes contain celestial bodies such as planets, suns, stars, moons, comets, supernovas, and many others. These planes could also be called the planes of the macro universe. Each planet, star, or moon, has its own independent Terrestrial Instinctive Operating Mind (TIOM). There are many versions of TIOM depending on the complexity of the celestial body. Some celestial bodies are inert and do not require an advanced TIOM but there are some advanced planets such as our own earth that contains complex life and consequently requires a more advanced TIOM. There are billions upon billions of celestial bodies of every description in our Physical Domain and each has its own version of TIOM.

Planes of the Galaxies – These are the planes that contain the galaxies of the universe. This could be called the planes of the super-macro universe. The galaxies of the universe are assigned to the various galactic planes depending on the evolutionary level of each galaxy. Galaxies like everything else are born; they grow up, mature, and eventually decay. This process takes place over billions of earth years but in absolute time it is only the blink of an eye.

Diagram 3
Planes of the Physical Domain

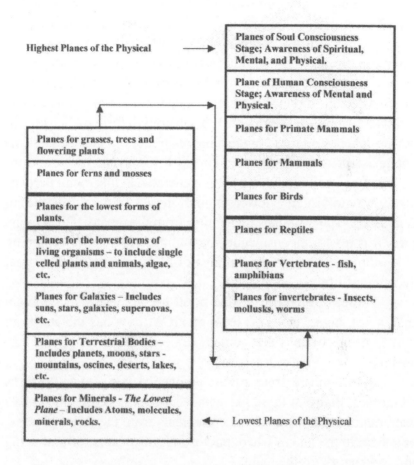

The Galactic Instinctive Operating Mind (GIOM) operates the functions of the Galaxies.

7.1.2 Planes of the Plant Kingdom

The Plant Kingdom is distributed throughout millions of planes. All plants are living organisms and consequently exist in higher planes than those of the inanimate materials. The Plant Kingdom represents a huge variety of plant life, and plant life is the oldest form of life on earth. Plant life is managed by PIOM, and each species of plant has its own Group Consciousnesses. The Plant Consciousness Stage covers a huge range of life, from the single-celled plants to the giant Sequoia trees. The following categories are broad representations of the various levels of plants, which correspond to level of evolution and occupy ever increasing levels of planes.

Planes of the Single-Celled Microbial Organisms - These planes contain the most primitive and lowest forms of life in the universe. The most primitive life first appeared on earth about 3.5 billion years ago and is still the most abundant life today. The most primitive form of PIOM and the most primitive form of Consciousness are inherent in these organisms. Awareness in these organisms is nonexistent, yet it is the start for all Consciousness. New Consciousnesses are continuously being generated at this point. Perhaps a likely way to visualize the creation of these Consciousnesses is to visualize a "vapory plume" that comes from the breath of the Divine Spirit. It is a very inconspicuous beginning, yet it is the beginning of Consciousness and the beginning of life. It is the start of the great adventure of life.

Planes of the multi-celled plant life – These planes contain fungi, yeast, algae, etc. This level of PIOM now has the ability to manage groups of cells that come together as multi-celled plant organisms.

Planes of the green plants – These planes contain the ferns and mosses. Plants at these planes have a root system, which takes water, minerals, and other food ingredients from the soil, and transports them to the leaves where photosynthesis occurs for producing food and energy for the plant.

Planes of grasses, trees, and flowering plants – These are the highest planes of the plant kingdom and contain the most advanced plants of the planet.

7.1.3 Planes of the Animal Kingdom

Animals belong to the Animal Consciousness Stage. Each animal species is managed by its own version of AIOM and each animal has its own independent AIOM. AIOM gradually becomes more advanced and therefore more powerful as it evolves upwards through the animal planes. Each animal species is a member of an Animal Group Consciousness and this is the start of the lowest level of awareness, which is mere sensation. Planes of the Animal Kingdom are more advanced than those of Plants Kingdom.

Planes of the Invertebrates –These planes contain insects, mollusks, earthworm, jellyfish, crabs, spiders, and many more. Invertebrate animals do not have an inner skeleton; instead they have an outer membrane or shell to protect them. Some of the animals in these planes are extremely abundant, for instance there are about 722,000 species of insects, and about 80,000 species of mollusk.

Planes of the Lower Vertebra Animals – These planes include the fish and amphibians. Animals at these planes have sophisticated organs, brains, and nervous systems.

Planes of the Reptiles – These planes contain the reptiles, which are more advanced than the fish of the previous planes.

Planes of the birds – These planes contain the birds, which are more advanced than the reptiles of the previous planes. There are some species of birds that are quite advanced and show signs of a rudimentary personality and a rudimentary reasoning mind. The more advanced birds can display emotions that have a high degree of fear and aggression as well as affection and romance. Many birds also show a high degree of parenting skills in the care of their young. All of these animal emotions become a part of the AIOM and are designed by nature to help the animal survive. The higher levels of birds mark the start of awareness of being physical. This means that they become aware that they can suffer pain or discomfort due to cold and hot, etc.

Planes of the Mammals – These planes contain the mammals and are more advanced than the planes of the birds. The higher mammals such as horses, elephants, dogs, cats, pigs, dolphins, whales and others show strong signs of a rudimentary Intellectual

Mind, which is the reasoning mind. Personality and emotional traits are also visible in all the advanced mammals.

Planes of the Primates – These planes contain the more advanced primates, which are more advanced than the planes of the mammals. Primates include monkeys, chimpanzees, orangutans, gorillas, other great apes, and humans. The great apes are the closest relatives to humans in terms of both physical and instinctive mind characteristics. The DNA of the great apes is about 98% identical to that of the human. DNA is a physical property rather than a mental property, but mind is what gives it its superb capabilities.

7.1.4 Planes of the Humans

The planes of the human are at the apex of planes of earth. Humans living on earth today belong to the Human Consciousness Stage and are governed by the HIOM. Human Consciousness and HIOM are enormously more advanced than the animal Group Consciousness and the AIOM. Humans have been a part of the Human Consciousness Stage since the beginning of our existence on earth and to this day, we are still a part of that same stage. We have advanced a great deal through our stage of Consciousness but we still have a ways to go. To show progress through our stage of Consciousness, the stage has been subdivided into sub-stages for clarity. The sub-stages are listed here in descending order of advancement for illustration, but the sub-stages will be described in more detail in Chapter 8. The sub-stages of the Human Consciousness Stage are as follows:

- Refined Human Sub-stage – This is the highest sub-stage.
- Civilized Human Sub-stage
- Intermediate Human Sub-stage
- Primitive Human Sub-stage
- Early Human Sub-stage
- New Human Sub-stage – This is the lowest sub-stage.

Planes of the New Human Sub-stage – These are the planes that contain the humans whose Consciousness has just recently made the transition from animal to human. In historical terms they would be equivalent to the most primitive Cave Dwellers, or Stone Age Man.

Planes of the Early Human Sub-Stage – These are the planes where human advancement is to the level of farming and live in relatively large social groups.

Planes of the Primitive Human Sub-stage – These are the planes where human advancement is to the level of the early Egyptian Civilizations and the early Biblical times. Generally these are still prehistoric times since writing is still in its infancy.

Planes of the Intermediate Human Sub-Stage – These are the planes where human advancement is to the level of the ancient Greek and Roman Civilizations. Generally this is the beginning to recorded history.

Planes of the Intermediate Human Sub-stage – These are the planes of human development equivalent to the period beginning with the start of Christianity and lasting through the discovery of the Americas by Christopher Columbus.

Planes of the Civilized Human Sub-stage – These are the planes of human development equivalent to the period of history starting with Columbus's discoveries and continuing through modern times.

Planes of the Refined Human Sub-stage – These are the planes of the most advanced humans of our present stage of Consciousness, which include the affluent, educated, professional, intellectual, religious, and compassionate people of today. The range of human development is huge and there is no precise demarcation between sub-stages; there is always a blending in from one sub-stage to the next. The descriptions above are generic for illustration purposes only.

7.1.5 Planes of the Soul Consciousness Stage

There are quite a few planes in the Physical Domain, which belong to the more advanced early levels Soul Consciousness Stage. The people at this stage of development are not numerous; they represent about 5% of the earth's population today. These people have the same level of HIOM as we do and are difficult to distinguish from the average. Their advancement is primarily due to their innate awareness of being Spiritual Beings.

7.1.6 Correlation of the Planes to Physical Evolution

The progression of the planes as depicted above is intended to represent the planes of the Physical Domain as they exist today. It is interesting, however, to note that the progression closely resembles the evolution of the cosmos and of life on earth as explained in Chapter 6. In other words, the first material things that were present in the universe after the "Big Bang" were the subatomic particles, and the simplest of atoms, which are represented by the planes at the bottom of Diagram 3. After that came the clustering of atoms to form stars, planets, moons, and all the other celestial bodies, and next came the galactic formations. All these correspond to the first three segments of planes as shown in Diagrams 3. The existence of inanimate matter was first, next came plant life, then animal life, and last came human life. This corresponds with the progression of the hierarchy of planes as shown by Diagram 3. The hierarchy of plane goes from the most primitive to the most advanced, and that corresponds to the way all things evolve, from bottom to top.

7.2 Astral Domain

The Astral Domain is similar in many ways to the Physical Domain. The two domains may be considered mirror images since they are similar, but they are also quite different from each other. The Astral Domain could be called an *Astral Universe* because it is as vast as what we normally call the *Physical Universe* but this book prefers the concept of one universe with multiple domains. In this concept all domains are interconnected and interdependent. Both

Evolution of Consciousness

the Physical and Astral Domains are superimposed on each other and generally occupy the same space. For instance, for every physical tree there is an astral tree; for every fish there is an astral fish; for every cow there is an astral cow; for every mountain and ocean there is an astral mountain and ocean; for every star, sun and planet there is an astral star, sun and planet, and for every human body there is an astral human body. The two domains are inextricably linked where the Physical Domain could not exist without the Astral Domain and the reverse is also true. The Astral Domain is made of astral matter and energy, which is similar to matter and energy of the Physical Domain, but the wave vibrations of the Astral Domain are much higher than those of the Physical Domain. The frequency vibrations of the Astral Domain are out of range of our human senses; therefore they are imperceptible to us. Mother Nature did not design our senses to perceive the Astral Domain because it has not been necessary for our development. Whenever we outgrow the capability of the Physical Domain to provide us with new experiences, then Mother Nature will see fit to upgrade our senses so we can continue to grow with the experiences of a whole new domain.

Astral matter is made of atoms and subatomic particles that are equivalent to physical counterparts, but the electrical polarity of astral atomic particles are opposite from those of the Physical Domain. Astral matter is similar but not the same thing as antimatter that our scientists detect in the high-energy particle accelerators. Antimatter shows up for extremely short durations in laboratory experiments, but because of its astral properties it switches to its normal higher frequency vibrations and it becomes undetectable again. Astral matter has weight (mass), volume, and occupies space, and it also responds to gravity. The Astral Domain has time and space, but the appearance is different. Distances between places appear closer, and time appears shorter.

Astral matter is the same thing that many present day scientists are calling *dark matter*. Physicists and astronomers of today believe that this so-called dark matter exists, but they cannot detect it with existing instrumentation. Some astronomers believe that as much as 70% of the matter in the universe is dark matter. It is called dark matter only because it does not emit visible light and therefore cannot be seen. Astronomers know that this unseen matter exists because of the gravitational effects that are detectable when studying the totality of matter of the Physical Domain. Studies indicate that the estimated total amount of physical matter in the universe does not correlate well with the estimated gravitational pull of that matter. They estimate that in order for the total gravity to balance out, there must be more matter in the universe, and that additional matter must be dark matter. In actuality this dark matter (astral matter) does emit light, but it belongs to the astral spectrum, which we cannot see. The astral spectrum is composed of much higher frequency than the electromagnetic spectrum and consequently has different properties and characteristics.

The Astral Domain has its own physics, chemistry, biology, and mathematics, which are similar to our own physical sciences. At present, only generalities with few specifics are available on the scientific nature of the Astral Domain. It is my opinion that eventually we will develop scientific knowledge on the Astral Domain. This is so because we have to understand more of the Astral Domain in order to understand the rest of the physics of the Physical Domain.

7.2.1 Planes of the Astral Domain

The Astral Domain, like all other domains in the universe, is divided into many planes and sub-planes, which are arranged in a hierarchical manner from lower to higher. The planes of the Astral Domain correspond one to one with those of the Physical Domain, that is, for every plane in the Physical Domain there is a corresponding plane in the Astral Domain. Diagram 3 for the Physical Domain also applies for the Astral Domain. The planes of both domains correspond to each other but they are not the same; they exist side by side but they actually have different functions. In all cases, the instinctive mind that operates the functions of an entity in one domain also operates the functions of that entity in the other domain. This applies whether the object is inanimate or whether it is a living organism.

7.2.2 The Inanimate Planes of the Astral Domain

The inanimate planes of the Astral Domain are the lowest planes of the Astral Domain and correspond to the same planes of the Physical Domain as follows:

- Astral Planes of Subatomic particles, atoms and molecular correspond to the physical planes of subatomic particles, atoms and molecular
- Astral planes of celestial bodies correspond to the physical planes of celestial bodies
- Astral planes of the galaxies correspond to the physical planes of the galaxies

The MIOM, TIOM, and GIOM are the instinctive minds for both the Physical as well as the Astral Domains.

7.2.3 The Plant Planes of the Astral Domain

The planes of the Plant Kingdom correspond to the same planes in the Physical Domain. One PIOM operates the functions of the plant in both domains simultaneously. In other words, the same PIOM operates the functions of a physical oak tree as well as the functions of the same astral oak tree. One Plant Group Consciousness animates each plant species in the Physical Domain as well as in the Astral Domain.

7.2.4 The Animal Planes of the Astral Domain

The planes of the Animal Kingdom in the Astral Domain correspond to the same planes in the Physical Domain. One AIOM operates the functions of each animal in both domains simultaneously. That is, one version of AIOM operates the function of the physical body of a dog as well as the astral body of the same dog.

One Animal Group Consciousness animates each animal species in the Physical Domain as well as in the Astral Domain. In other words, there is one Dog Group Consciousness for all dogs in both domains.

7.2.5 The Human Planes of the Astral Domain

The human is enormously more advanced than the animal, in view of the fact that each human has an independent Consciousness, an Intellectual Mind, a Complex Personality, and HIOM with a Trainable Mind that the animal does not have. The astral planes of the human are perfectly correlated to the physical planes of the human. In essence there are two human bodies (one for each domain) but only one person with all its human qualities. Note that Diagram 3 shows only one rectangular block for each stage of Human Consciousness, but in affect each block represents the various sub-stages of human and each sub-stage contains thousands of planes.

7.3 Vital Energy

Energy of many different types exists in every domain of the universe. Each type has its critical function within its domain. *Vital Energy* is the name designated for energy that belongs to the Astral Domain, and it is essential to life on earth. Life on earth would not be possible, as we know it, without Vital Energy of the Astral Domain. The fundamental energy source for both the Physical Domain and the Astral Domain is nuclear energy that is generated at the core of the sun. The sun is the primary source of energy for all life on earth, and is transmitted to earth in the form of sunlight. Sunlight is part the electromagnetic spectrum in the Physical Domain, as well as a part of the astral spectrum in the Astral Domain. Plants use sunlight to produce food and energy for themselves. This plant food and energy is produced by using the process of photosynthesis in combination with water, carbon dioxide, and other minerals from the earth. In turn, animals eat plants to obtain their food substance and energy. In the Physical Domain, energy in living animals is measured in calories or body heat. Heat is emitted from the body in the form of infrared (IR) radiation. The electromagnetic spectrum is the primary transmitter or radiator of energy of the Physical Domain, and likewise the astral spectrum is the primary transmitter or radiator of energy in the Astral Domain.

Evolution of Consciousness

Vital energy is the energy that animates life and is just as critical to life as caloric energy is in the Physical Domain. Vital energy is present in great quantities everywhere on earth and throughout the universe. It is found in the air we breathe, in the water we drink, and in the food we eat. Fresh clean air in the great outdoors contains more vital energy than stagnant air in a closed room; clear clean spring water in a flowing stream has more vital energy than water in a stagnant pond; raw living foods from fresh fruits and vegetables have more vital energy than dead foods such as cooked meats or canned fruits and vegetables.

7.3.1 Vital Energy in Humans

The total energy in humans is composed of physical caloric energy, astral vital energy, mental energy and Spiritual energy. All forms of energy are essential for the human to live, and this composition may be called the *personal energy* of each individual. The *personal energy* of each individual is surrounded by the ocean of *Universal Energy*. There is a thin membrane of energy that separates the personal energy from the universal ocean of energy. The universal ocean of energy is limitless, and it is free for us to use at all times. Energy is always flowing in and out of our personal energy reservoir. It flows in and out in a multitude of ways, and it is important to try to keep the body charged up with energy as much as possible at all times.

It is natural, and obvious, for young people to have boundless energy. Most of this energy, which is essential to keeping us alive and healthy, is vital energy that belongs to the Astral Domain. Young people with an abundance of vital energy have abundant health and exude energy in all their activities. Older folks, on the other hand, have more trouble retaining vital energy and consequently are more susceptible to getting sick or looking unhealthy or sluggish through most of their daily activities.

Vital energy is present throughout the body and in every cell. There are seven energy concentrations called *vital energy centers* (chakra in eastern philosophy) in the human body. Clairvoyants can see these energy centers as little vortexes that spin very fast in those people with high energy. People that are sluggish, tired, sick, or old have low energy and the vortexes spin correspondingly slower. The general locations of these seven vital energy centers in the human body are as follows:

- At the top of the skull; also called the crown.
- At the forehead and between the brows; also called the third eye.
- At the front of the throat.
- Over the heart.
- At the navel.
- Over the spleen.
- At the base of the spine.

7.3.2 Human Auras

Vital Energy radiates out of the body in all directions and creates an aura around the human body. Clairvoyants are gifted people that can see wavelengths that belong to the astral spectrum and can indeed see human auras. In the Physical Domain, human bodies radiate heat in the infrared (IR) segment of the spectrum, and form an IR aura around the body. These IR auras can be detected with infrared sensing equipment, which the military uses as night vision equipment. There is no equipment that can be used to see the astral aura. Only a few gifted clairvoyants can see that.

The astral aura can give much information about a person's health and/or mental disposition. It is said that Edger Casey, the famed physic, could see auras of people when he was an adolescent and thought that it was normal for everyone to see auras as he could. It was not until he was a young adult that he realized that he had a special gift that few people have. Mother Nature can and will eventually furnish us all with the gift of seeing in the Astral Domain. Remember that Mother Nature gives the owl the sense to see in the dark, the hound dog has the ability to track people long distances by the sense of smell, and the geese get the ability to navigate half-way around the earth on their migratory journey. By the same token, Mother Nature will give all of us the appropriate senses whenever She determines that we are ready for them. That will become necessary in order to expand our horizons and give us the potential for greater experiences and growth.

7.4 Thought and Thought Dynamics in the Astral Domain

Thoughts are very real things; they are as real as electricity, light, or air. Thoughts are created by the Intellectual Mind and then transferred to the brain via the HIOM. The brain is not a thinker it is a doer; it gets its instructions to do things from the mind. There is an interface mechanism that translates instructions from the nonphysical mind to the physical brain. Once the brain has the translated instructions it causes physical actions to occur.

Thoughts are mind that has been energized with astral energy; consequently they belong to the Astral Domain. Humans generate thoughts using their Intellectual Mind, and then pass these thoughts to the physical and astral brain through the HIOM. The amount of energy added to the thought determines the strength of the thought, and this strength of thought depends on the desire, will power, and determination of the person creating the thought. We generally generate thousands of thoughts, and all thoughts have varying degrees of strength. For instance, thoughts that are generated by people with strong convictions, strong desire, and strong will to accomplish have more power than thoughts that are generated by people in a daydream state with little or no conviction. Thoughts are made of waves that belong to the astral spectrum, and consequently are a form of astral energy. Thoughts radiate out from the source much as light radiates from a light bulb. Each person can be said to be a beacon radiating thought, be it either good or bad, or strong or weak. Thought, with the power of desire and will, can be focused and directed to other people or places, or it can linger around like a cloud. Because thoughts are made of astral energy, some clairvoyant people can see them. The reports from such people are that thought waves look like small vaporous clouds, of multi-colorations and hues that drift in the astral atmosphere. Clairvoyants can see thoughts with their wide range of colors in a similar manner as they can see auras. A physic person that can see thought clouds could also learn to interpret the cloud's coloration and determine if the thoughts have either good or evil connotation.

Thoughts are influenced greatly by the natural "Law of Attractions" (like attracts like) so that thoughts of similar type and character will tend to come together or congregate in the same general location. Some thoughts may linger around for a long time at the location where they were originated, and may even linger long after the person left the area or died. Thoughts of similar subject matter may also congregation at great distances from their origin. Most people put little force into their thoughts, and such involuntary thoughts have little motion or affect, however some of these low energy thoughts may be given motion by attraction from other like-minded people. Like-minded thinkers tend to attract thoughts of a similar nature. A thinker in deep thought will attract thoughts of a similar character and force them to migrate, coalesce, or band together in his or her own mind.

The thought-atmosphere of a community is the average composite thought of the people living in the community. Localities, cities, or nations consequently have their own characteristic thought-atmosphere, which is peculiar to the average combined thought of its people. When a person moves to a new location, that person will be exposed to the thought-atmosphere of the new community and will eventually add to, or absorb the prevailing thought patterns of the new location. In this way a person may sink or rise to the level of the prevailing thought-atmosphere of the community.

Places such as buildings, homes, colleges, churches, and etc. take on the general thought atmosphere of the people that inhabit them. For example, a home may have a sunny, cheerful, and happy atmosphere, while another home may have a cold, sad, repugnant, and unhappy atmosphere. The same principle applies to individuals; some people may project cheerfulness, a bright sunny disposition, or courage, while others may project feeling of gloom, distrust, uneasiness, or discord.

A place where a violent crime has been committed will carry an unpleasant atmosphere, attributed to the vile thoughts of hate and fear from both the criminal and the victim. The atmosphere of a maximum security prison is horrifying to a sensitive person not accustomed to such a place. The atmosphere of vice, alcohol, drugs, prostitution, is suffocating to one of high mental traits.

A university may be said to be a liberal school, a conservative school, a party school, a high-minded school, etc. The general ambience of the university depends on the caliber of students that are attracted to the university. Students of like-mind tend to be attracted by the predominant thought-atmosphere and general environment of the university.

7.4.1 Effects of Good or Evil Thoughts

High-grade thoughts as well as low-grade thoughts often occupy the same astral space. High-grade thoughts are good thoughts that are defined as positive thoughts of love, kindness, compassion, and forgiveness, while low-grade thoughts are evil thoughts defined as negative thoughts of hate, fear, cruelty, and blame. Good thoughts are made of higher frequency wave vibrations within the astral spectrum, while evil thoughts are made of lower frequency vibrations. As a rule of thumb, good thoughts are charged with a higher degree of energy potential, while bad thoughts carry relatively less energy potential, and as a consequence of that principle good thoughts generally override evil thoughts. This is a simplistic statement since there are actually many variables and many factors involved. An important factor that imparts strength to the thought is the intensity of the desire and will behind the thought.

The following is a list of guidelines that apply to the behavior dynamics of good and evil thoughts. This assumes that the intensity of desire and will is equal throughout.

- Each individual draws to himself or herself thoughts that correspond to his or her way of thinking. A person with good thoughts will attract good thoughts from the astral

atmosphere, while a person with bad thoughts will attract bad thoughts.
- Good thoughts are made available equally to good and bad people. Normally good people benefit from these good thoughts while bad people reject them due to ignorance.
- A person of ill will can radiate evil thoughts, which will be attracted by other evil thinking people, and those thoughts will be detrimental to both the sender and the receiver. A person with good thoughts will reject evil thoughts and will not be harmed by them. Evil thoughts will literally bounce off a high-minded person as if a reflective shield was built around that person, and the harm will instead fall on the sender. The reason this is so is because the higher frequency thoughts contain more power than the lower frequency thoughts.
- A prayer for protection is always recommended to reinforce the good thought environment and to reject low grade thoughts. These guidelines imply that on the long run "good will" always triumphs over wickedness.

7.4.2 Problem Solving

If a person is working on a problem that is baffling and he or she cannot readily solve it; it behooves that person to assume a receptive attitude by holding thoughts along the same lines as the problem. It is likely that after a time when the person stops thinking of the problem, the solution will flash in his or her mind as if by magic. Some of the world's best philosophers, engineers, scientists, writers, speakers, and inventors have used this wealth of information that is available to all from the astral atmosphere. Most of these thinkers stumbled on the source of information subconsciously without ever realizing where the information came from. The Astral Domain is full of unexpressed thoughts in every field of endeavor, and they are there just waiting to be utilized.

A person can, in a similar way, draw to himself or herself strong helpful thoughts that will aid in overcoming fits of depression or discouragement. There is an enormous amount of stored up energy in the thought world, and anyone who needs it may draw that which he or she requires. It is simply a matter of demanding, that which is free for your use, as you need it. This demanding can be done through the use of prayer, meditation, or simple contemplation.

7.5 Human Astral Body

The human "astral body" is the equivalent counterpart of the human "physical body". It is similar in appearance to the physical but there are important differences. The astral body is made of astral matter whose nature is made of waves that vibrate at very high frequencies, much higher than its physical brother. The astral body is superimposed and intimately connected to the physical body throughout the lifetime of the person. The normal condition is for the two bodies to be superimposed on each other, but there are instances in which they separate. Three types of separation are as follows:

- **Sleep Separation** – The two bodies separate during sleeping, or fainting, or coma.
- **Astral Projection or Out of Body Experience** –In Astral Projection a person can separate the astral body from the physical body under mentally controlled conditions.
- **Near Death Experience** – These occurs on special situation when a person is in the process of dying, usually due to an accident, but is saved from death at the last moment, often by doctors in a hospital.
- **The Process of Death** – This is similar to the Near Death Experience, but in this case there is permanent separation of the physical and astral bodies and the Consciousness goes on to the Spiritual Domain 1.

A brief explanation of each process follows.

7.5.1 Sleep Separation

During waking hours the physical and astral bodies are together (interpenetrating each other). When you go to sleep, the astral body may slip out of the physical body during deep sleep but normally it stays close by. The same sort of things happens when a person faints, or goes into a coma, or blackout. General anesthesia, which is commonly used by doctors to put a person to sleep for surgery, does the same thing. The reason anesthesia works is that the feeling of pain is in the astral body and not in the physical body as commonly considered. Therefore, when the astral body slips out of the physical body due to the affect of anesthesia the doctor can perform operations on the physical body with no pain. When the separation occurs there is a connection between the two bodies through what is called a *"silver cord"*. If anesthesia is not applied correctly the *silver cord* may break and the person may die in the operating table. The physical body remains alive only as long as the silver cord remains connected, if or when, the silver cord brakes the physical body dies.

7.5.2 Astral Projection or Out of Body Experience

Ordinarily conscious separation of the two bodies is difficult, but it is possible for some people that are of high Spiritual development. People that have reached a high degree of Spiritual attainment can accomplish "Astral Projection or Out of Body Experience" at will. It can also occur to some people accidentally under certain circumstances, but it is not the normal method. The advanced person can accomplish the separation, usually by lying down in a quiet place and meditating, and then by force of will, it is possible to separate the astral body from the physical body. For people that can do this separation, the physical body remains behind in a restful environment, while the astral body goes off on long journeys. During the separation, the connection between the two bodies is maintained through the long slender *silvery cord,* which if broken, causes the physical body to die.

The astral body is invisible to the ordinary person because it is made of high frequency vibrations that are imperceptible to most of us. It can float through the astral atmosphere and pass through solid physical objects at will. The person merely has to think of where he or she wants to go and then feel as if flying effortlessly through the air. Gravity, or distance, or weather is not a hindrance. It is possible to go through solid physical objects because of the much higher frequency composition of the astral body. It is similar to x-rays that can pass through solid objects, such as flesh or bone, due to the high frequency composition of x-rays.

When a person goes on a journey in the astral body, that person takes with him or her all the conscious faculties and is fully aware of everything that is going on. He or she can later remember all that transpired during the trip. The person leaves the physical body behind but takes along all the component parts to include the astral body, most of vital energy, most of the HIOM, Personality Mind, Intellectual Mind, and Soul-Consciousness. What stays behind is enough HIOM and vital energy to sustain the physical body.

While on an astral trip the person has its astral senses activated, and can see the astral world with the corresponding astral senses. The astral senses are sharper than the physical senses, therefore, a person can see better like the owl, hear better like the bet, or smell better like the hound dog. With the astral senses a person can see people as ordinary people, and can also see their astral auras. The color of the auras can give an indication of the health and mental disposition of the people encountered. The new astral senses give the person the potential for new experiences that were previously impossible, and now this person will have the tools for further growth.

In the astral journey the person can also see thought forms in the shape of vapory thought clouds. Thoughts, like the auras, are made of astral energy, the difference being that they are in different segments of the astral spectrum and therefore have different qualities and characteristics. As the person travels in its astral body, he or she can see places, especially around large cities, where there are large concentrations of both high-grade and low-grade thought forms. These thoughts are seen as thought-clouds arising from the city like great clouds of smoke or vapor. High-grade thought-clouds are made of astral colors that depict joy, cheer, or high-minded endeavors, while low-grade thought-clouds project astral colors of gloom, sadness, fear, and vile expressions. The person that can see in the Astral Domain is called clairvoyant, and a person that can hear in the Astral Domain is called clairaudient. At the end of the journey, the conscious mind merely has to think that it is ready to get back to the physical body and the process of reuniting with the physical body takes place almost immediately.

7.5.3 The Near Death Experience

Near death experience usually occurs to a person who comes close to dying due to an accident such as suffocation, heart attack, or head injury but the person is saved before real damage is done to the physical/astral brain or other vital organ. This experience is involuntary but normal, and it is called near death because the person is revived before serious physical/astral damage occurs. Often doctors at a hospital bring the person back to life before real death actually sets in. Several books have been written on this subject of near death experiences. These books are usually about reports from people that have actually had the experience. There have also been several excellent documentaries on television on the same subject. The reports of near death experiences are similar from one person to the next but they are never identical; there are always subtle and unique variations from person to person. Culture, age, gender, nationality, religion, or economic status seems to have an affect on the experience. In almost all cases they claim that their lives change for the good after the experience. Most feel that it was a positive experience that makes them more aware of life.

Most people that go through a *near death experience* report the following sequence of events. Note that the people reporting do not normally know about the astral body. When they give details of their experiences they are referring to their awareness of the events.

1. At the time of death, the astral body separate from the physical body and it seems to float over the physical body. The astral body hovers over and around the vicinity where the physical body is lying, and the person is able to see his or her lifeless physical body lying there. The person can see and hear everything that is going on, but is unable to communicate with other people in the room.
2. After hovering over the physical body for a short time, the dying person then feels as if he or she is being sucked through a tunnel, and at the end of the tunnel there is a bright white light.
3. The reporting people always have trouble describing the white light precisely; they do not seem to have the right words to express the sight or feeling behind the light. Most describe the light with statements such as "a brilliant white light", "overwhelming feeling of tranquility", "wonderful feeling of joy and peace", or "incredibly warm and loving light".
4. The next step generally is that many of these people seem to be welcomed by loved ones that have passed away previously. The welcoming committee may be organized of parents, family members, or friends that have passed away a long time before. They are there to give comfort, but if need be they may also inform them that it is not time for them to die. Most people are given the choice of going back if the physical body is still in good enough shape to sustain life. The reports that we get are of course from those who chose to come back. Those who choose to stay, go through the death experience, and do not come back to report their experience to us.
5. Most people that have the near death experience go only as far as seeing the bright light, and then return to their body, but some manage to go further in the regular death process. Some report that they go through what is called the, *"life's review"* in which they see their entire past life replayed before them as in a 3-deminsional movie. This unedited movie

vividly flashes through the mind's eye in seconds before actual death takes place. The movie contains every detail of the person's life, including all the good and the bad. This review gives the person the opportunity to appraise the past life and to make a judgment upon him or herself. In some cases the experience can be breathtakingly awesome or it may be terrifying depending on the type of life lived. The review is designed to allow the person to remember past experiences and to learn from past mistakes, which were mostly due to ignorance.
6. When a person goes back to the physical body, it means that the silver cord did not break and the process did not go far enough to real death. At that time all the normal aches and pains come back. The choice to go back is given only if the physical body is repairable enough to continue to live. The affect that this experience has on people's lives is almost always mystical and glorious. Many feel that they have gone through a Spiritual experience, and they feel compelled to try to live exemplary lives in the future.

7.5.4 The Process of Death

Death is a mystery to many people of the Western World. The Christian faith is vague about the subject of death. It is shrouded in obscurity and ambiguity and it causes many to fear death because of the apprehension of the unknown. Christianity teaches that there is some kind of life after death, and that it is associated with either going to heaven or hell for the rest of eternity. Where you spend that eternal life depends to a great extent whether you have been good or bad according to the rules of the church in the present life. Many people consequently tend to fear death and to many the subject is unmentionable. There appears to be paranoia about death and it seems that your eternal life hinges on what happens at the time of death.

Death is merely a transition to another phase of life. The terms *transition* or *passing away* are perhaps better suited because

Evolution of Consciousness

they do not have the finality implication that the term death has. The process of death takes place in three domains. It takes place in the Physical Domain, the Astral Domain, and in Spiritual Domain 1(SD1). This section of the book will explain the part of the process of death that takes place in the Physical and Astral Domain. The next chapter will explain the part of the process of death that takes place in the SD1. Consciousnesses that reside in the SD2 and SD3 are far to advanced and do not have to reincarnate into bodies in order to evolve, therefore they do not experience the death process any more.

7.5.5 The Process of Death in the Physical Domain

There are two basic types of death, normally called natural death or accidental death. Natural death is when a person dies in his or her sleep, in bed due to natural causes such as old age or old age disease, while accidental death is a sudden and unexpected death. The process of death begins when, (1) the body looses its vital energy and (2) the internal vital organs begin to fail. These two processes exacerbate or aggravate each other. For instance, if vital energy is low, then, the vital organs may start to fail, or if vital organs start to fail, then the vital energy is diminished.

The main function of the HIOM is to keep the body alive and it works feverishly to keep it alive as long as it has the energy and organs to work with. As the body begins to fail, the HIOM does not fail, weaken, or decline during this process; it only loses the tools that it has to work with and becomes ineffective as the situation deteriorates. A computer analogy that can help to illustrate this is that of the Operating System of a computer. If the hardware of the computer fails, it causes the OS to function incorrectly, but the OS is still good and complete. The OS does not change or deteriorate when the hardware components fail. The OS stops working only because it loses its hardware components with which to carry on its job. By the same token the HIOM is OK but cannot function if the physical body fails.

Accidental death is more sudden, but the principles are still the same. In accidental death a vital organ is damaged beyond repair, and the HIOM cannot repair the damage. The vital energy departs from the body in rapid succession, and the body dies.

At the time of death the physical energy departs the physical body and the body goes limp. The astral body separates from the physical body and carries with it some astral vital energy. The astral body then hovers over the physical body for some time, much as it is described by reports of people that have had a near death experience. The astral body continues to remain connected to the physical body through the *silver cord* until the energy is drained from the physical body. When the silver cord breaks the physical body is dead.

At the time of death all the higher component parts of the human leaves the physical body, and without energy the physical body immediately starts to decompose. All the living cells of the body also lose their energy and dies in rapid succession. When the cells die they quickly breakup into their individual molecules, and atoms and go their separate ways to become a part of the natural organic environment. These materials, atoms, molecules, and minerals go their separate ways, and take with them sufficient vital energy and MIOM to operate on their own until the next cycle begins. Eventually these materials will start a new cycle by becoming a part of other living plants, and then eventually a part of a living animal or even another human.

7.5.6 The Process of Death in the Astral Domain

At the time of death, the Consciousness discards the physical body but takes with it the astral body, the astral energy, the HIOM, the Personality, and the Intellectual Mind. For a natural death the astral body survives for a short time (normally from 30 minutes to as much 24 hours) after the death of the physical body.

There are many abnormal variations to death. One fairly common abnormality (less than 2% of all deaths) is when the astral body survives for days or even months after the physical body dies. In some extreme circumstances the astral body may survive years after the death of the physical body and this is called a *ghost*. This situation is not normal but can occur due to strong psychological or emotional problems that the person had prior to its physical death. Many times excessive grief by family members causes the dying person much consternation and anxiety and may cause an abnormal departure. Extreme amounts of hate, bitterness, grievances, or horror may keep the astral body from passing on.

If the death is normal, which most deaths are, the astral body will linger around the physical body for a short time and then the Consciousness discards the astral body along with the rest of the HIOM. At this separation the *"second silver cord"* breaks. The second silver cord is what connects the astral body to the Consciousness. At the time when the astral body dies, it loses the remainder of the vital energy and becomes an empty shell. The empty shell of the astral body is similar to a dead human corps; there is nothing there. The astral body begins to disintegrate and decomposes immediately in a similar fashion as the physical body. The empty shell of the astral body tends to drift to where the remains of the physical body are laid to rest, and they decompose together.

The Divine Light – Soon after the break of the first silver cord the Consciousness goes through what is described as a *tunnel* for lack of a better term. This is the same tunnel that people with near death experiences described. At the end of the tunnel is the "white light" which project peacefulness, joy, calmness, and love. This light is of a *Spiritual Nature*; it is not a part of the electromagnetic spectrum as some people may be inclined to think. This bright light is of the Spiritual spectrum, and that is the reason it has those special qualities of peace and love. This light is *Divine Intelligence* and comes in part to provide comfort during the transition. The reason this wondrous light is necessary is because most people are unknowing and therefore tend to be fearful of death. Our own social traditions, and religious teachings and beliefs cause many of us to fear death. The comforting light immediately relaxes those fears by projecting love, not with words but with wonderful feelings of kindness, calmness, and assurance. The process can then proceed with a greater degree of ease. After the dying person is relaxed he or she is greeted by loved ones such as parents, other family members, and friends that have long since departed. It must be emphasized that death is not something to be afraid of, it is not the end of life; it is only the beginning of the next part of the cycle of life.

The Life's Review - The next step after the Divine Light is the viewing of *"Life's Review"*. Only a few people that have had the *near death experience* go through this viewing, because most do not get that far. In real death experience all people go through this viewing. In the *life's review* the entire life of the person, from infancy to old age, passes before the mental vision as if watching a 3-dimensional movie. As the life's story rapidly plays before the mind, many things are made clear and plain to the Consciousness. The Consciousness has a chance to reflect, and understand the meaning of the earth life just completed because it can see it as a whole. This *life's review* is like a dream intended to leave deep memory impressions, which are designed to allow learning from past mistakes. Life's mistakes are mostly due to ignorance, and when that ignorance is eliminated the Consciousness grows and evolves. This movie is unedited and contains every detail of the person's life, including all the good and the bad. If the person lived a terrible life the review is terrible; if the person lived an exemplary life the review is good.

The Slumber Period - The next step after the life's review is for the Consciousness to start getting sleepy, but may last for several days or weeks; the length of time varies a great deal depending on the individual's development. When the Consciousness awakens it immediately goes to a plane in Spiritual Domain 1 (SD1). The plane that the Consciousness goes to is the most appropriate for it according to its level of development. By this time the Consciousness only carries the Intellectual Mind and the Personality Mind.

The HIOM is not destroyed when the Consciousness discards it; instead the HIOM goes into an archival system where it is kept until it is needed again. The archival system of nature keeps an immense amount of data. It keeps such information as the life reviews of all lifetimes for all people, and more. Nature's archival system is much more efficient than anything we have on earth today.

Awakening is SD1 after the slumber period is discussed in the next chapter.

Chapter 8

The Spiritual Domains

The Spiritual Domains are not physical places or locations; they are states of being in which Consciousnesses reside; they could not exist without Consciousnesses. Spiritual Domain 1 (SD1) is the lowest in the hierarchy, SD2 is the next higher, and SD3 is the highest. The level of the Consciousnesses that resides in them is what gives them their ranking. All three Spiritual Domains have prescribed planes, which may be portrayed as planes of attainment for Consciousnesses. The less advanced Consciousnesses inhabit the lower planes of a domain while the more advanced Consciousnesses inhabit the higher planes. Less advanced and more advanced refers to the degree of evolutionary attainment of the Consciousnesses.

The Spiritual Domains are of a "mental character" and are as real as real can be. At our stage of development we tend to think that realness only applies to physical things, but that is not so. SD1 has equal reality as the Physical Domain because both domains are inhabited by the same stage of Consciousnesses, and consequently they both have the same degree of relativeness. SD2 is residence to the more advanced Soul Consciousness Stage and the Cosmic Consciousness Stage of humans, which makes this domain more real than SD1. That is, SD2 is farther along than SD1 on the "scale of relativeness". SD3 is the most real of all domains because it is occupied by humans of the Spiritual Consciousness Stage that is the most advanced stage of Consciousness and therefore is approaching absoluteness. Spiritual Domains only have human Consciousnesses in them. There are no Plant Group Consciousnesses or Animal Group Consciousnesses in the Spiritual Domains.

8.1 Spiritual Domain 1

SD1 is a "Mental State of Existence" and only people of the Human Consciousness Stage reside there. SD1 is where we live when we are not living an earth life. Our life in SD1 is the other part of our cycle of life. This is also called by such names as the other side, the after life, happy hunting ground, heaven, and many others. SD1 provides a "state" for us to rest following the previous earth life; it is also a state to prepare for the next earth life. This is all part of our ongoing development.

SD1 is closely associated with the Physical Domain and the Astral Domain because it is our stage of Consciousness that goes to live in SD1 when not living an earth life. The higher stages of Consciousness such as the Cosmic Consciousness Stage and the Spiritual Consciousness Stage do not incarnate into the physical due to their advanced status. Consciousnesses at the Cosmic Consciousness Stage and Spiritual Consciousness Stage continue their evolution at SD2 and SD3 respectively without needing the Physical Domain.

The level of the plane in SD1 where the Consciousness resides is proportional to the level of unfoldment of that Consciousness. Consciousnesses are distributed through all the different planes according to their degree of development. The less advanced Consciousness live in the lower planes of SD1 and the more advanced Consciousnesses live in the higher plane of SD1. The level of the plane (attainment) has nothing to do with the equality of the people living in them. In the eyes of God all humans are equal regardless of their evolutionary progress; that is, all Consciousnesses are equal under Universal Law.

It should be emphasized that all people are equal. In the words of Thomas Jefferson, in the *Declaration of Independence,* "all men are created equal" holds true on earth as well as in every other part of the universe. Every person though, is in his or her own individual and unique level of development; independent and yet interdependent with everyone else. Another important principle is that every individual lives at his or her best and most suitable evolutionary level and plane. All humans are equal regardless of where on the "scale of evolution" they are. It must be cleared, that it is offensive to Mother Nature for the more advanced people to take advantage of the less advanced people. Taking advantage of the less developed has been the rule of mankind for eons of time on earth, and has consequently led to a lot of misery at every level of our society. It has been like that for countless ages and it is still prevalent. The prognostication from the advanced Guides is that this will change in due time. Bigotry is a serious ill in our society that needs correction. Bigotry is almost nonexistent at the Soul Consciousness Stage.

It should be noted that the words of Thomas Jefferson were beautiful and far-reaching, but it can be construed that at the time of the writing the words "all men are created equal" meant that only white men that owned property were equal. It did not include women, slaves or poor people. There has been a great change in our philosophy since that time; this is good example of how far our Evolution of Consciousness has progressed since the days of Thomas Jefferson in 1776.

8.1.1 Evidence of Spiritual Domains

There is no way to prove the existence of SD1 using physical means since SD1 is nonphysical and our senses were designed to perceive only the physical. We cannot explain what the senses cannot perceive. Agnostics and skeptics demand physical proof but there is none and no attempt is made here since no amount of explanation can satisfy their needs.

It is recommended that the reader think upon the subject and experiment. It is hoped that this material will provide the planting of the seed, which will later grow and bear fruit. As the person grows, develops, and evolves forward, many of the old ideas will eventually no longer make sense and will be discarded. As the old ideas are discarded they will be replaced with new and more advanced ideas.

Mother Nature's future plan for us is to augment our present five senses with more powerful senses that can provide us with perception of the presently unseen domains. This should occur whenever she knows that we are advanced enough to merit the enhancement. These additional senses will be provided, not to prove to the skeptics that there is more than just the physical world, but because we will need them in order to continue our forward progress. Some of the early new senses will be analogous to those that clairvoyants now possess. The new higher senses will make the humans of the future seem as superhuman to us today. They will have the ability to see the colors of the Astral Domain, to interpret colors in human auras, to read thought, and to travel in the astral. The new senses will probably be in line with the principles of the Aquarian Age, which are supposed to increase spirituality and harmony in people's lives starting sometime in the 21st Century (our present century).

Our vocabulary does not have words to explain the unseen domain; consequently in order to really understand it we must be able to see it first hand. This book tries to explain the unseen by using metaphors of physical experiences, which approximates the unseen. Some examples or analogies to explain how words cannot explain the sensory experiences are as follows.

- Suppose a person is born colorblind and has never seen the color "red" and you try to explain the nature of the color red in your own vocabulary. You may say red is crimson, ruby, cherry, scarlet, but those are also colors that the blind person cannot see. The blind person has no color reference to start from, so using another color to try to explain the color red will be useless. You might change your approach and use words like flashy, dazzling, bright, beautiful, or any other description but to no avail. The American Heritage Dictionary defines the color red as, "The hue of the long-wave end of the visible spectrum, evoked in the human observer by radiant

energy with wavelengths of approximately 630 to 750 nanometers; any of a group of colors that may vary in lightness and saturation and whose hue resembles that of blood. **b.** A pigment or dye having a red hue". Again this definition does not help the colorblind person visualize what the color red really looks like.
- By the same token, how do you explain in words to a deaf person from birth, what the sound from a symphony orchestra is like? Words cannot describe what only the senses can provide.
- How do you describe the taste of cinnamon to a person who has never tasted cinnamon? Or how do you describe the smell of camphor to a person who cannot smell? Our vocabulary was not made to describe what only the senses could explain. The senses in combination with the mind and Consciousness are what give us the awareness of the physical world around us. Our present senses were not designed to see the Spiritual world; words cannot easily convey what the senses can experience.

8.2 The Slumber Plane and the Awakening

In the previous chapter we discussed the process of the physical body dying in the Physical Domain and the astral body dying in the Astral Domain. We also mentioned that the Consciousness goes into a deep slumber period before awakening in SD1. The slumber period takes place in a specially assigned "Slumber Plane" of SD1; it is a plane where all Consciousnesses go to rest for a short duration after living a hectic life on earth and before waking up in SD1. This time of rest can vary anywhere from a few weeks to as long as a few months of earth time. The Slumber Planes are very important from a Spiritual standpoint. The slumber period is Spiritual requirement and the planes are heavily supervised by advanced Consciousnesses that come down from higher realms. The Slumber Planes are seriously guarded in order to keep high integrity and security. It is important to keeping disturbances or intruders out. Only authorized

Consciousnesses of impeccable standing are allowed in to supervise the uprightness of the Slumber Plane.

Before describing the awakening process of the Consciousness in SD1 we should review the process of dying in the Physical and Astral Domains, which were discussed in a previous chapter. The process of dying is similar for all people but it is never identical. The differences come from the fact that each person has their individual experiences and beliefs that makes for the variations. This review is formatted as a step-by-step list of events.

Step1 - The physical body loses all energy, dies, and is discarded.

Step 2 - The astral body separates from the physical body.

Step 3 - The Consciousness goes through a tunnel and finds a wonderful white light at the end.

Step 4 - The Consciousness views the "Life review".

Step 5 - The silver cord breaks and the Consciousness discards the astral body and the HIOM.

Step 6 - The Consciousness goes into a deep slumber period for several days, weeks or months.

Step 7- The Consciousness awakens from the slumber and goes directly to a plane within SD1 that is most suitable for its evolutionary attainment.

8.2.1 Awakening in SD1

Upon awakening in SD1, the Consciousness goes directly to the plane that is most appropriate and suitable for its level of development. This is equivalent to the Consciousness resurrecting in SD1 with awareness of the previous life on earth. All Consciousnesses of equal levels of evolution go to the same equivalent plane. The less advanced Consciousnesses go the lower planes and the more advanced Consciousnesses go to higher planes. There is no *"Saint Peter"* at the proverbial *"pearly gates"* to direct the Consciousness where to go. The signature frequency of a Consciousness is automatically attracted by the signature frequency of the proper plane with a high degree of accuracy. A Consciousness cannot go to a plane that is higher than what he or she is capable of attaining. The suitability of the plane for a Consciousness is directly dependent on the level of evolutionary advancement of Consciousness. This applies at every level of the universal hierarchy. Note that often in this book the terms Consciousness and person are used interchangeably. This is so because they are one and the same. It is always the same Consciousness whether living an earth life or living in SD1.

At the point of awakening in SD1 the Consciousness notices many things that are different and feels flabbergasted to some degree but it is aware and conscious of everything. The Consciousness can recognize many things that are familiar and yet many more that are different and strange. In SD1 the Consciousness still has the same thoughts and values that it had on earth prior to so-called death. Death makes absolutely no change on the memory of the Consciousness; it does not get any smarter or dumber as a result of death on earth. However there are many differences in SD1 that will take some getting used to. For instance, SD1 does not have time or space, or mass and inertia, which is difficult to understand at first, but most of the differences are always for the better and it takes a relatively short time to get used to them.

8.2.2 Examples of Differences upon Awakening in SD1

Every person is different with different experiences, different sets of beliefs, and different levels of development; therefore death is somewhat different for each individual. The most important thing to remember is that the mental attitude and beliefs of the person are what determines the first impression of SD1 for that person. The differences in mental attitudes range from the advanced person to the savage and from the normal to the abnormal. Abnormal conditions include those who commit suicide or those who die of drug overdose. There are big differences between those who believe in a Divine Being and those who are atheists. Also there is a big difference between those who are gentle and kind, and those who are mean and cruel. There are those who preach hellfire and brimstone and have strong beliefs in the devil. On the other hand, there are those who study metaphysics and know what to expect. In affect, we are all different and each one of us presents a unique circumstance depending on our own mental attitudes and belief system. Examples of a few of these situations follow.

Those who commit suicide due to inner turmoil have a hard time when they first awaken in SD1 because they find that death is not final. They find that there is life after death and the problems that caused the inner turmoil in the first place are still there. The problems do not go away because of death, thus suicide is not the answer; it is actually easier to resolve those problems while still living on earth. A person does not have the right to take its own life anymore than it has the right to take the life of another person. The inner torment continues in SD1 for a certain period of time but eventually the person sees the stupidity of it and sooner or later repents. After repentance the Consciousness can in due course proceed again on the path of development.

An atheist who is a total unbeliever in a Divine Being or the hereafter or life beyond the grave is totally unprepared and will not try to adjust to the unexpected situation. The atheist usually has a rude awakening in SD1, after insisting all his life that there is nothing beyond this earth life. At first he is astonished and later resentful and may blame it on some kind of hallucination or figment of his imagination. Gradually and grudgingly however, he or she begins to accept the error of the belief and eventually tries to learn the truth of the unexpected condition.

There are those who believe in the Judeo-Christian Bible literally and who believe in hell, hellfire and the devil, and who believe that they lived exemplary moral lives on earth. For instance, a clergyman who preaches on the scourge of hell and the devil probably also believes that he is righteous, moral, virtuous, upright, and honorable and believes that he is going straight to heaven and will sit next to God with his entourage of Angels. When it does not happen as he envisioned it, he tends to believe that those he meets in SD1 are lost Souls and tends to try to preach to them there. But nothing seems to make sense and eventually he begins to listen to those advanced Consciousness who are there to help him. They explain that basically he is a good person but has been preaching a wrong doctrine. Eventually he begins to contemplate upon his mistaken principles and eventually realizes his blunders and makes amends by opening his eyes to the truth. He then is open to take classes in the true philosophy and make headway in his own path to unfoldment.

Many who die of accidental death or in war suffer great shock because it is hard for them to accept having to suddenly give up whatever it was that they were doing on earth. Usually they are young with high aspirations and to suddenly lose that is heartbreaking. Of course, there are always advanced Consciousnesses who immediately become available after the accidental death to explain the new situation and to try to ease the level of shock. Eventually they get over it and continue in a normal fashion.

Drug addicts who die of overdose have an especially difficult time in SD1. Drugs are chemicals that affect the physical brain, which in turn affects the mind. There are no chemical drugs in SD1 and the urge for the drug does not go away. There is no way to experience the kind of high of illicit drugs when living in SD1. The addict has to go through a terrible period of drying out that is at least equally as severe or worst than drying out in earth life. The experience leaves them weak for a long period of time but the craving for the drug slowly subsides. Eventually the need for drugs fades into the background and the person is able to function and live a semi-normal life again. The effect of the addiction can continue for several lifetimes on earth. Only the addict can conquer the scourge through self-discipline. A bad habit never goes away completely because it is indelibly engraved into the subconscious database. The best we can do is to remove it from the foreground of the mind through self-discipline, and with time it will slowly fade away into the background of the subconscious and thus become harmless.

When an Aborigine or a savage dies with no exposure to the thinking of modern Western Civilization, he or she normally has little trouble adjusting to the ways of SD1. An Aborigine who lived in primitive villages will probably continue to live in SD1 as if nothing changed. He will probably still hunt and fish and live of the land.

Of course in SD1 this is strictly a mental process but in most cases it is difficult to tell the difference.

The people who have the easiest transition are those who have studied the principles of metaphysics and know what to expect. They waste little time getting on track and moving forward. Normally when they first get there, they do all the preliminaries of visiting with old friends and relatives that are already in SD1 and then they get acquainted with the new environment. The knowledge that everything is mental in SD1 gives them the upper hand. They know that if they think in terms of beauty, goodness, and joy, they encounter those qualities in SD1. With the proper attitude things such as flowers and trees have more vivid colors, and the colors of the sky and its sunsets are more vibrant. They find that the beauty and the joy in SD1 surpass that of earth. Once that phase is over the Consciousness goes to school to learn more philosophy on the higher-minded dogma. The variety of examples is to numerous to recount therefore only the gist of the concept is provided here.

Evolution of Consciousness

Diagram 4
Planes of the Spiritual Domains

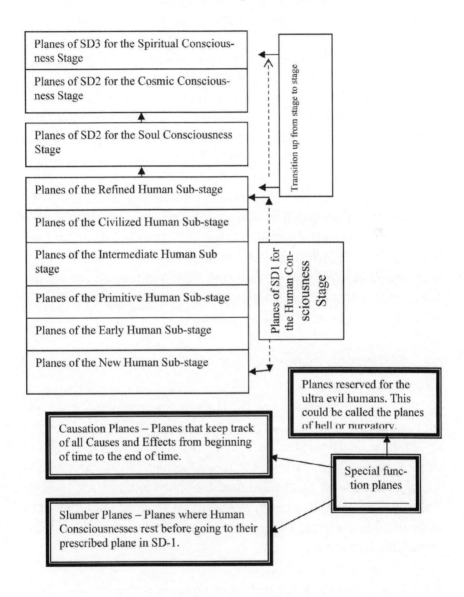

8.3 Planes of Spiritual Domain 1

Refer to the Diagram 4, which depicts the arrangement of the planes for all the Spiritual Domains. Each block represents millions or perhaps billons of planes. The least advanced planes are at the bottom of the diagram, and the more advanced planes are at the top. Progress in evolution is always made from lower to higher. It must be noted that this graphic is a representation of the planes of the Spiritual Domains for illustration purposes only; there is no true scale to the diagram. All the planes coexist in the same space because waves of different frequencies can coexist in the same space. This is the same principle of radio and TV broadcasting which transmits many signals in the same space without interfering with each other. The analogy of several flashlights of different colors shining on a wall without interfering applies here also. This principle is the same in all Spiritual Domains. In SD1 for instance, each plane is made of many frequencies of the Spiritual Spectrum and they all exist in the same space without interfering. In actuality "frequency dimensions" rather than space represent these planes.

8.3.1 SD1 has no Planes for Plants, Animals or Inanimate Matter

Spiritual Domains are for Human Consciousness only. Inanimate matter does not have Consciousness therefore there are no planes for that purpose in SD1. Plants and animals have Group Consciousness but their lives are for living earth lives only; consequently there are no planes assigned for plants and animals in SD1. Plants and animals do exist in the Spiritual Domains but they only exist as mental images (thought forms) of the Human Consciousnesses that live there.

8.3.2 Planes for the Human Consciousness Stage

SD1 is exclusively for people that belong to the Humans Consciousness Stage. Our level of Human Consciousnesses either lives in the Physical/Astral Domain or in SD1, depending of the cycle of

Evolution of Consciousness

life that we are in. Each of the three domains has equivalent planes, but note that they are only equivalent and not the same. There are millions of planes in SD1 and each plane has its own "Signature Frequency" that attracts Consciousnesses of corresponding "Signature Frequency". The "Signature Frequency" of the Consciousness is the primary indicator of the relative advancement of that Consciousness. This principle is what allows each plane of SD1 to contain Consciousness of equal development. The "Signature Frequency" is an accurate indicator of the development of the Consciousness; it is one of Nature's critical indictors and it does not make mistakes. The Signature Frequency is the result of a complex algorithm that factors in all experiences, accomplishments, emotional stability, Spiritual development and all else of the Consciousness. It depicts the location of the Consciousness on the "Scale of Evolution of Consciousness".

This book subdivides The Human Consciousness Stage into six sub-stages for the purpose of clarification. These six sub-stages are explained in more detail in the next chapter. They are merely listed here to illustrate the principle of human planes in SD1. The list is in descending order and is followed by a short explanation.

- Refined Human Sub-stage
- Civilized Human Sub-stage
- Intermediate Human Sub-stage
- Primitive Human Sub-stage
- Early Human Sub-stage
- New Human Sub-stage

In accordance with Diagram 4, the lowest planes for Human Consciousness are shown at the bottom of the diagram, and are occupied by the lowest New Human Sub-stage. The diagram shows only one block for each human sub-stage but in actuality each block represents thousands or perhaps millions of planes and sub-planes. For instance, the New Human Sub-stage occupies thousands of planes, and it advances slowly from plane to higher plane.

8.3.3 Causation Planes

There are special planes in SD1 that are dedicated to perform certain functions. The "Slumber Planes" were discussed above. Another set of planes are the "Causation Planes" which performs a very complex function, and it is impossible to give an in depth description for lack of knowledge of the subject. Generally, the Causation Plane keeps track of all causes and effects that occur throughout the universe. The records of "Causes and Effects" are permanent and they register all causes and all effects from the beginning of time to the end of time. Nothing is ever forgotten and nothing is ever lost. The activities of the Causation Plane work closely with the Universal Laws – Cause and Effect, Relativity, and Evolution.

8.3.4 Barrier above each Plane

Consciousnesses of the same level of development occupy equivalent planes of attainment. "The signature frequencies" of the Consciousness has to match the "signature frequency" of the plane. Therefore, the plane that best matches the Consciousness is the plane, which provides the most comfort and happiness for the Consciousness while living in SD1. When the Consciousness reaches that most appropriate and highest plane to live in, it is not aware that there are planes higher than its own. Consequently the Consciousness always tends to thinks that it lives at the summit of existence, it does not know and is not aware that there are other Consciousnesses that are more advanced. The Consciousness is aware of planes below it, but is not aware of planes above it. A Consciousness can go visit friends and loved ones at equal or lower planes, but it cannot go to higher planes than its own plane. There is a barrier between planes, which does not allow Consciousnesses to penetrate, if they are not qualified. This is a "universal principle" that is unbending, and there is no arbitration.

8.3.5 Analogies to Explain Barriers between Planes

Fish in the Water - One analogy to explain the principle of barriers between planes is that of a fish in the water. The fish is limited to living in the water and the surface of the water presents a barrier that keeps the fish in the water. The fish, if it had awareness, could not and would not be aware that there is existence above the barrier. A barrier restricts the Consciousness in its plane as the surface of the water restricts the fish in the water.

Wire Screen Sorters - Another analogy to clarify the idea of barriers further is that of wire screens used for sorting stones. Suppose that there are three wire screens with different wire grids from coarse grid to fine grid. The coarse grid screen stops the large stones from going through but allows the smaller stones to go through, and each screen thereafter does the same for smaller and smaller stones. The end result is that the stones are graded according to size. The principle is similar when sorting Human Consciousnesses through the different planes. Of course, Human Consciousnesses are not sorted by size, but by the level of attainment. The Human Consciousness cannot go above the barrier of its own plane, but it can go to any plane beneath its own plane.

8.3.6 Living in the Higher Planes of SD1

Life at the upper planes of SD1 is much easier than life on earth. Some people may think that they died and went to Heaven because it is so pleasurable there, but do not be confused because it is not Heaven. Heaven is much grander, it is a state of wondrous beauty, love, joy and ecstasy; it is true bliss and we are all headed in that direction in due time. In the mean time, SD1 is a great place (state or dimension) to be in. It is great for instance, because we do not have to worry about having to work for a living, or worry about food and shelter, or worry about crime such as theft, or murder, or death. In SD1 you are always young and healthy with no need to worry about sickness. You are free to pursue the activities that you like most, and you can have as many friends of like mind as you want.

8.3.7 Living in the Lower Planes of SD1

The lower planes of SD1 are those where the more undeveloped but normal people of earth live. This refers to people who are good people but are just starting up the ladder of development. This includes the planes for sub-stages of the New Human, the Early Human, and the Primitive Human. For them living in the lower planes of SD1 is not much different than living on earth. The Consciousnesses that live in the lower planes of SD1 are very materialistic and can only be happy in earth living. So, when they pass on to SD1 they mentally shape all their surroundings to look like earth since that is what they know best and that is what they are the most comfortable with. Many may only vaguely recognize that there is any difference. For instance, if they like to hunt and fish on earth, they most likely will want to hunt and fish in SD1. If on earth they like to befriend others to tell stories, or play music, or play dominos, or whatever, then that is probably what they do in SD1. They still have huge advantages over living on earth in that they do not get old, they do not get sick, they do not get hungry, they do not have to fear wild animals or harsh weather, etc. SD1 is their "happy hunting grounds".

8.3.8 The Planes of Hell or Purgatory

People, who live with low-grade thoughts such as hatred, deceit, sensuality, etc., and live lives of crime and evil on earth, will gravitate to low-grade planes in SD1. These planes are reserved for the ultra-evil Consciousnesses who in earth-life committed some of the most horrific crimes on humanity. This does not apply to the average person who on occasion does something wrong because of peer pressure, or severe temptation, or because they carry out wrong deeds due to ignorance. It only applies to those people whose entire lives are totally and deliberately bad.

These planes where evil people reside are separated from the regular planes of SD1 because of their horrific nature. They are repugnant and hideous for normal people, and should be avoided at all cost. The influence that these planes could have on the average normal person could be devastating. These planes are what are commonly called "Hell" by some "fire and brimstone" preachers in our society. Dante envisioned fire in Hell; well, there is no physical fire there but it can feel every bit as hot if the mental attitude imagines fire. The types of people that reside in these planes are criminals of the worst kind, such as those who belong in maximum-security prisons. They are serial rapists and killers, those who torture or torment others without remorse, and even take pleasure in their evil deeds. Many of these people are violent psychopaths. They are madmen, homicidal maniacs, savage terrorists, genocidal murderers, diabolic zealots, or others who commit horrific crimes on humanity. The people who live in these planes of hell are not only those who commit and execute the crimes, but it can also include those who think along the same lines. Remember that thought is just as real as the action; evil thought is always the precursor to evil action.

These planes are full of ugliness, which matches the ugly disposition of the people who live there. The ugly imagination in their mind is what creates this ugly environment. It is so gruesome that it would make a normal person vomit in disgust if he or she were to enter this environment by accident. Yet there are some people in earth-life who try to experiment with "black magic" or "satanic cults" who have a propensity to stumble into these planes and sense its ugly effects. Some of these people are young adults trying to experiment for the sake of adventure and excitement. The pity of it is that they are ignorant and unaware of the plausible results. Entering into these planes for whatever reason normally has devastating consequences on their lives. It is highly recommended to refrain from practicing "black magic", "devil worshiping" or any other similar endeavor.

These planes of "Hell" are not intended to be vindictive punishment for delinquent people. They are intended for the purpose of giving the delinquents the opportunity to see the error of their ways and to make amends as a consequence. Nature does not put them there for revenge or punishment; she is not a vindictive being. This condition does not last forever. The delinquent can eventually move on to better conditions when he or she has truly recognized the error and begins to make the proper changes. These evil people are given every opportunity to repent, and mend their ways. They are given every chance to get out of the hell planes and proceed in their evolutionary path in a normal fashion. As long as there is some ray of hope for them to give up their evil ways, Nature is willing to give them a second chance.

There are a few evil Consciousnesses who are so evil and so far gone, that Nature loses all hope for them to ever come out of their evil state. In such case Nature eliminates them altogether and forever. This is the only instance known, where Nature completely vanishes, destroys, annihilates and eradicates a Consciousness. Otherwise all Human Consciousnesses are indestructible and eternal. They all go to bigger and better existences.

The torture and suffering received in these planes of hell are all in the form of mental images in the mind of the perpetrator, but they are as real as real can be. For example, an evil person in earth-life who believes that sins will be punishable by fire and brimstone will not be disappointed when he or she reaches SD1. The person's beliefs supply the necessary environment, and the person's conscience condemns the person to the punishment of his or her beliefs. The person will receive the pain and suffering commensurate with the degree of foul deed committed. The result of the suffering provides the appropriate and valuable disciplinary lesson, and in turn produces remorse and a desire to mend the ways. The experience also carries forward in dim recollections to the next incarnation. There are many degrees of "hells" which are dependent on the various scales of dogmatic beliefs or religious influence. Each hell has the punishment, which is best suited to exert a deterring influence over the person in his or her next life.

Each Consciousness creates its own heaven or hell – for neither has any objective existence. Neither heaven nor hell has a physical counterpart; they are both created in the mind of the Consciousness. The heaven or hell of each person is the result of the person's belief system, and the cause and effect created in earth-life. This mental creation may seem strange to us now but the creation is nonetheless very real to the Consciousness. There is nothing in earth-life that is more real than these experiences. In essence, each person is his or her own lawgiver; each person dispenses his or her own glory or gloom; each person also dispenses his or her own reward or punishment. The amount of suffering is proportional to the degree of the criminal deed performed, and on the other hand the amount of joy is proportional to the degree of good deeds performed.

Again, there is no one overseeing or directing this process, there is no St. Peter at the Pearly Gates, it is all done automatically by natural law. When we change our mental attitudes and thought patterns, we change our mental frequency vibrations and in so doing we are attracted to the appropriate plane. The "Law of Attractions" applies here, as it applies all over the universe.

8.4 Characteristics of Life in SD1

SD1 is a "mental state of existence" consequently thought is of paramount importance. Your thoughts are what dictate your lifestyle and your environment in SD1. Philosophical viewpoints, religious beliefs, cultural surroundings, political affiliation, educational background, and professional career all have an important bearing on your thought pattern and thereby influence the way you live in SD1. Most things that you are familiar with in earth life you will encounter in SD1 and it will be just as real if not more so.

You can use the analogy of a dream in earth life to roughly visualize what SD1 is like. For instance, a dream in earth life can seem very real; it can seem as real as real can be, and it is only perceived as a dream when the person wakes up the following morning. A dream can have real people, real things, and real places that are completely recognizable. Everything has shape and form and seems as real as physical things. The same applies to SD1 except that mental things in SD1 are more real than physical things on earth. SD1 is not a place for rest, relaxation, and idleness it is a "mental state of existence" where you remain mentally active and you continue to grow and evolve. The following are a few of the phenomena that are encountered in SD1, some of which are similar to Earth life but most are quite different.

8.4.1 Diversity

The diversity of SD1 is enormous when compared to the diversity of all that we know of in our planet earth. We normally think of earth as being a diverse place to live in. For instance, we know we have great variety in music, art, theater and movies, and in books and literature; there are wide varieties of cultures, philosophies, and religions, around the world; there are also huge varieties of plants and animals and natural wonders. Now, try to imagine SD1 as having much more diversity than earth in every category. This is so because it is a mental phenomenon, which grows with every Consciousness that brings in new experiences. It is difficult for our intellect to comprehend that this kind of diversity can exist in a non-physical environment; it is outside our range of understanding. There are planes and things within those planes that are unimaginable to us. The discussion in this book is a simplified glimpse of what actually exists in SD1. It will all be available for us to examine in due time.

8.4.2 Old Age

There are no old people in SD1. If you happen to die of old age and sickness on earth, upon awakening in SD1 you will no longer be old and sick. Instead you will now be at the prime of life all over again. You will see your body as vigorous as in youth and also as real and substantial as back in the physical world. The prime of life is considered to be equivalent to age thirty-five on earth. Generally this is the age at which we reach mental maturity on earth, yet we still have plenty of physical endurance, vigor and strength. Children that happen to die on earth at a young age will go to SD1 at the age of their death on earth, and continue to grow and mature until they reach their prime of life. Their growth in SD1 is similar to growth on earth except that it is somewhat faster than earth time. In SD1 we do not have to worry about having to work for our food and shelter. We never have to worry about going hungry, or of having to have car insurance, or house or health insurance. There is no need to worry about having to pay VISA or taxes; there is no worry of poverty and pestilence. So, you can see there are many advantages to living in SD1.

8.4.3 Dimensions

Our physical world is a three-dimensional world, that is, there are three dimensions of space; but now, since Einstein's revelation, we can add one dimension of time. SD1 has four spatial dimensions, which appear, as space but there is no time. It is difficult for us to get used to the new reality when we first wake up in SD1, but we soon grow accustomed to the new situation. The four dimensional world cannot be explained in words because we do not have a comparable condition in the physical world. The amazing thing is that once we are in SD1 we develop the required senses to perceive the new conditions. Once the person gets accustomed to the new conditions, a vast set of new experiences is opened for assimilation. As an analogy a three-dimensional tree in the great outdoors is more revealing than a picture of a two-dimensional tree on canvas.

SD1 does not have physical time or space, or mass or inertia as we know it on earth. But, in a different sort of way, time is a dimension in SD1 since time is a component of frequency. This is so for the reason that frequency is measured in terms of cycles per unit of time. So in actuality, Consciousnesses travel in planes of frequency.

8.4.4 Family

Family life is nonexistent in SD1, at least not as we know it. Family relationships are for earth life only. Family members on earth are only friends in SD1. A mother on earth, for instance, is no longer a mother in SD1; she is now a close and loving friend who served the role of a mother on earth. Your brother, sister, aunt, uncle, or father on earth, are now only friends that are highly revered. You can visit ex-family members that reside in SD1 at any time, especially those with whom you had loving bonds. There will be many meetings with ex-family members but no large family reunions are likely since they would be to big and cumbersome. If there were family members on earth with whom you had animosity, you probably do not wish to meet with them. You only visit those people with whom you are most comfortable with, or whose thoughts are most closely aligned to yours. If you had acrimony towards a person in earth life, it is best to remove whatever resentment there is by forgiving the person while still living on earth because it can continue to create pain in SD1 if you don't forgive and make amends.

There are no marriages in SD1 and no babies are ever born there. People with a loving marriage on earth will most likely meet again in SD1, and if mutually satisfied they may continue to stay together as companions for an expended period. People that had bad marriages on earth will probably never meet again in SD1. If they do happen to meet there and they had animosities on earth, those animosities will continue to persist there. The best thing to do in that case is to break it up as soon as possible and allow time to eventually heal those animosities. The healing should be done consciously, if not the Law of Cause and Effect will eventually prevail, and in due course both parties will see their error and correct it.

Close friends as well as professional or religious acquaintances on earth would also be close friends that are highly valued in SD1. Wherever there is an emotional link of love on earth, a meeting is likely to occur in SD1. Enemies on earth will probably never meet in SD1 but if they do it will be painful and should be avoided.

8.4.5 Clear Thought

The mechanism of thought processes and thought manifestations in SD1 are different from that of earth-life, yet there are also similarities. Thinking in earth-life is much more cumbersome. In SD1, if you can think of something clearly you can have it almost immediately. If you cannot get what you want, it is because you have not learned to think clearly enough. That same principle applies in earth-life as well as in SD1 life. If you can think clearly and forcefully on what you want, you can have it whether you are in earth-life or in SD1 life. The difference is that it is much easier in SD1. Thought forms in earth life encompass the Intellectual Mind, the HIOM, the physical brain, and the astral brain; whereas, thought forms in SD1 only encompass the Intellectual Mind.

In SD1, thoughts manifest into objects much faster than they do on earth, primarily because earth has time, space, matter and inertia, while SD1 does not. Space, time, and inertia are handicaps that we have to live with on earth, but they are absolutely essential for our growth at our stage of development. Planet earth is our school; it was designed with these parameters for our evolutionary process.

Thoughts create objects in SD1, which are made of mental substances that are just as real to us as material objects are on earth. A clear thought will manifest an object almost immediately, but when no longer needed this object also decays or melts away quickly. For example, a suit of cloths that is no longer worn by a person on earth may hang in the closet for years, gradually deteriorating, and finally at some point it is thrown away. In SD1, if the suit is not worn or needed any more, it quickly decomposes into its original substances. For this reason SD1 always seems clean and free of trash.

8.4.6 Desire, Will Power, and Strength of Thoughts

Thoughts are what make things happen in any and all domains of the universe. Thoughts are mind substance that has been given energy by force of WILL. The more energy that is imparted to the thought by the WILL, the more powerful the thought. All manifestations, be it physical manifestations, mental manifestations, or emotional manifestations are the result of thoughts. The Consciousness commands the Intellectual Mind to think and the Intellectual Mind generates the thought. The Consciousness first has to have a DESIRE for something. The Intellectual Mind then forms the thought and imparts a certain amount of WILL POWER to the thought, and the amount of WILL POWER is dependent on the strength of the DESIRE. The stronger the DESIRE of the Consciousness, the more powerful is the thought. It is extremely important to control your thoughts because positive thoughts bring about positive results, and negative thoughts bring about negative results. You have the power to control your thoughts.

8.4.7 Thought communication

Probably one of the most startling differences encountered in SD1 is that communication between people is done by thought and not by words. There is no such thing as language barriers in SD1. Thoughts are read directly by all, and therefore there is no room for secrets, deception, hypocrisy, or pretense. It is difficult at first for some people to adjust to this condition, and many suffer great discomfort for a period of time, but adjustments are usually made in due time.

8.4.8 Money, Knowledge, and Education

There is no money in SD1 and none is needed since there is nothing to buy. The reason money is not necessary is because you can immediately have whatever you want by clear thinking. It is well to understand that material wealth gained while living on earth cannot be taken to SD1. It is also well to understand that intellectual knowledge gained while on earth is utilized very effectively in SD1. Professionals and intellectuals such as teachers, college professors, artists, musicians, scientists, engineers, philosophers, writers and the like are in their element when living in SD1 because intellectual pursuits are appropriate there. SD1 provides a fitting environment for intellectual endeavor and encourages scholarship.

8.4.9 Companionship in the Spiritual Domains

The Human Consciousness is a social being which requires companionship, friendship, love and affection whether in earth-life or in any of the Spiritual Domains. The Consciousness instinctively craves the friendship and love of its fellow humans. The bonding created by people in earth-life is extended to the Spiritual Domains, but the higher the plane of existence the closer the bonding of companionship, fulfillment, and love. Earth bound relationships of love such as between a man and a woman, mother and child, brothers and sisters, casual and professional friendships will flourish in the Spiritual Domains. Because there are no physical bodies in the way, the Consciousnesses can attain much closer relationships than are possible in earth-life.

The longing and dreams for close friendships or close relationships may have been difficult for some people to obtain on earth, but in SD1 quality companionships are easy to find. Close bonding between Consciousnesses is now the norm. The higher the scale is, the higher the degree of love and the higher the degree of bliss. Love is easy to find in SD1; it is also joyful, fun and natural.

8.5 Principle of Reincarnation

8.5.1 Evidence of Reincarnation

There is no physical proof that can be given on the principle of reincarnation, but there is abundant rationale that can be used as evidence. For instance, simple logic should imply that it is impossible to be as smart as we are in our present earth-life without the advantage of having lived many previous lives on earth; this is so, given that each lifetime contributes to our present accumulation of knowledge. There is no way that one single lifetime can make us as smart as we are today. Evidence can be surmised if you consider such geniuses as Amadeus Mozart who started playing musical instruments and also composing musical masterpieces at a very young age. By logical deduction it can be inferred that no one can become that smart in the span of only one lifetime; getting smart doesn't just happen, it has to be earned. It should stand to reason that Mozart had to have acquired that knowledge of music over a period of several prior lifetimes. Similar reasoning can be construed of Albert Einstein, Clark Maxwell, Henry Ford, Thomas Edison, and others in all ages and in all walks of life.

Reincarnation is a concept that is poorly understood and difficult for people of the Western World to accept, and it is not the intent of this book to try to convince any one to accept the principle. For those that do not accept the principle, it is suggested that they read through the concept as presented in this book for the purpose of familiarization, and later examine the logic, experiment, and form opinions.

The Judeo-Christian Bible does not make strong references to life in SD1 or to reincarnation. The main reason for this is that several ecumenical church council meetings were held in the 4th and 5th centuries AD to try to resolve nagging questions of the early church dogmas. These councils actually modified the early Bible to suit the theological beliefs of the church clergy of the day. The first of these councils was convened and presided over by the Roman emperor Constantine in Constantinople in the year 325 AD. This first meeting was called the First Council of Nicaea and was attended by 318 bishops of the early church. There were a total of eight such meetings over the preceding years. It was at these ecumenical council meetings that they established such doctrine as the Church Creed, the duality of Jesus as being both Human and Divine, and they also settled the matter of the Trinity (The Father, the Son, and the Holly Ghost). Along with the many doctrines created by these councils, they also modified the Bible in some significant ways; among these is the deletion of references to the principle of reincarnation. The principle of reincarnation was unfamiliar, baffling, and unsupported by the logic of the time; in essence the early church fathers did not believe in reincarnation, so they removed most of the references. Consequently the principle of reincarnation was never taught by the Catholic Church and was basically forgotten.

Western religion generally does not support the existence of the concept and in some cases it is regarded as taboo. The western mindset has created many false views and myths of reincarnation and contributed to its suppression or even censorship. One of these false myths for instance is that humans can be reborn into the bodies of animals, which is completely untrue. Evolution is always forward, and never backward. In the east, reincarnation is central and fundamental to the religious belief system of Hinduism and Buddhism.

8.5.2 Reincarnation

The primary purpose of God's creation of the universe is to provide a means for the Evolution of Consciousness. Reincarnation is one of the main principles that contribute to that end. The universe, as we know it, would not be possible without the principle of reincarnation.

After living in SD1 for an extended period of time the Consciousness starts to develop a craving to live another earth life. The craving comes from an instinctive urge to continue to develop those things that can only be developed in an earth environment. The craving is instinctively developed by the Consciousness and is encouraged by the Soul and Spirit. Soon after the craving develops, a search for earth parents begins. The search is complex; it includes the search for the right environment that is conducive to growth, and the search for the right family members that promote a measure of stability for growth. It also requires the right time in history that offers potential for growth. The Consciousness tends to look for familiarity of people, things and environment. Each individual Consciousness is different from all others and each person has its own unique circumstance; in general the "Law of Attractions" and the level of attainment of the Consciousness play an important part in the selection process. Usually the child is born to parents that are at a slightly higher plane of attainment than the child, but as the child grows up the child tends to catch up and even surpass the parents, often by a wide margin.

Re-birth into the Physical Domain is a gradual process. It starts with declining activity in SD1 and a desire to rest or sleep on the part of the Consciousness. With the natural process of living in SD1 nearing its close, the Consciousness starts to feel a tiredness or weariness, and instinctively becomes sleepy and longs for a rest. By this time the Consciousness feels that for the most part it has completed its desires, and ambitions that it set out to accomplish in SD1. Now it starts thinking in terms of going back to another earth-life to fulfill other new desires that were built up during its life in SD1. The thinking and desire for another earth-life gradually builds up, and at the same time the longing for a rest also builds up. It is like reaching the end of a long, tiring, but wonderful journey, and now it is time for a well-earned rest.

8.5.3 Final Preparation and Influences for Rebirth

The Consciousness may have lived in SD1 for a few earth-years, or a hundred earth-years, or a thousand earth-years depending on its degree of development. The less developed Consciousness spends less time in SD1, whereas the more developed Consciousness spends more time there. When it becomes time to go, the Consciousness feels that its work in SD1 is done; it is now tired and ready to pass on to the next phase and its new life. There are many influences that attract the Consciousness to reincarnate in another earth-life. Some of these influences are as follows:

- The influence may be a desire to put into practices the advancements that were gained during its life in SD1. During its life in SD1 the Consciousness grew and evolved. It most likely dropped many undesirable characteristics due to the feeling of remorse and repentance, and by the same token it probably acquired many desirable characteristics during that same period. The Consciousness changes during its lifetime in SD1 just as it changes during its lifetime on earth.
- The influence may be an attraction to the material world to live out desires, which it never finished in a previous earth-life. The "Law of Cause and Effect" and the "Law of Attractions" guide the Consciousness to its new life on earth.
- The influence may be a desire or wish to be with some loved one who has recently reincarnating and is now living an earth life.
- The influence may be a desire to go back to some adventurous circumstance or surrounding environment.
- The influence may be the desire to live with certain parents and family.

- Each Consciousness goes to where it belongs by reason of what it is. There is no one directing where it should go. Everything is decided by the Consciousness in a more or less subconscious or instinctive mode, and you can rest assured that everything is fair, just, and equitable. The law of "Cause and Effect" is in full operation every step of the way. The law is fair and applies equally to all; there is no favoritism; ignorance has no barring in this case. We are all on the same evolutionary path.
- In actuality there are thousands of desires, requirements, and needs that drive the Consciousness to develop the inducement to go back to earth-life.

8.5.4 Dying in SD1

All these desires accumulate enough momentum to cause the Consciousness to die in SD1. Dying in SD1 is in some ways similar to dying in the earth-life. Of course, no one actually dies a permanent death. It is always better to call it a transition from one cycle of life to the next. At this time the Consciousness again goes into a deep sleep or slumber, which is similar to the previous deep sleep, after the death on earth. A period of rest (sleep, slumber, repose) between major milestones is always required by Nature.

Just before going into its deep-sleep the Consciousness has a fast "life re-run" of its completed life in SD1. This is similar to the "life replay" of the previous earth-life. This "life re-run" in SD1 can be compared to a movie on video tape or DVD; it allows the Consciousness to see its complete life in SD1, and gives it the opportunity to review, digest, and assimilate all its past thoughts, desires, activities, deeds, undertakings, and accomplishments. Remember that life in SD1 is full of activity that contributes to the unfoldment of the Consciousness. People in low levels of attainment tend to do less and thereby achieve less than people in higher levels of attainment. People in the higher levels tend to accelerate their rate of growth due to higher activity and a higher willingness to achieve.

8.5.5 Final Rebirth and the New Life on Earth

The Consciousness starts to awaken from its deep-sleep or slumber period while the mother back on earth is still carrying the baby in her womb. The Consciousness waits, and from time-to-times, it checks on the pregnant mother to ensure that everything is going well. The Consciousness does not enter the baby's body until the baby is out of the womb and takes its first breath, or in some cases shortly thereafter.

The Consciousness does not fully awaken after entering the baby's body. It continues its slumber and only gradually awakens as the baby grows through its first three or four years of infancy and early childhood. These early formative years are crucial to the developing child. It is important, from a psychological standpoint that the child is brought up in a well balanced, loving and caring environment; otherwise there will probably be a period of emotional upheaval in later years. The child normally does not consciously remember these early years, because the Consciousness is not fully awake yet, but every activity is recorded in its subconscious. Many of the early childhood experiences will, in later years, come to fruition as either positive or negative behavior.

As mentioned, the normal awakening of the Consciousness is slow during the child's infancy and early childhood. The rate of awakening coincides with the rate of growth of the child's physical brain. There are some abnormal conditions, such as when the awakening is faster than normal, in which case the child may be thought to be a prodigy or genius. Another extreme is when the awakening is slower than normal, and the child is considered mentally challenged. The typical awakening usually goes through age 18 of the young person.

There is a mechanism in Nature, which does not allow the adult person to remember the events of the childhood years; it is a similar mechanism of Nature that does not allow the average person to remember (consciously) the experiences of past lives. These experiences are indelibly recorded in the subconscious database but cannot be remembered. Even though not remembered, these experiences have a profound impact on the life of the individuals. The

Evolution of Consciousness

experiences of past lives are what give each person the intellectual abilities, the character, the personality, and the overall disposition. The blocking mechanism is in place to avoid direct interference of past lives with the present life. Only the overall essence of the experiences carries forward. Nature will only allow whatever is necessary for the evolutionary process to be carried forward; superfluous or unnecessary baggage is not carried as conscious remembrance. All the detailed information of all previous lives is archived for later use as the Consciousness evolves. The archival capacity of Nature is infinite; it is infinitely greater than any of the present day hi-tech data storage devices, such as DVD. Because the higher domains of the universe operate at much higher frequencies, their broadband capabilities for data storage, transmission, and assimilation are much greater than our physical broadband capabilities.

8.6 Spiritual Domain 2

Spiritual Domain 2 (SD2) is the next higher domain after Spiritual Domain 1, and is occupied by people who are more advanced than we are. SD2 is divided into two halves where the bottom half is for people of the Soul-Consciousness Stage, and the top half is for the more advanced Cosmic Consciousness Stage. These people are high achievers who work hard in pursuit of self-advancement in the philosophy of the Divine. SD2, like all other domains in the universe, is made up of many hierarchical planes. The higher the planes are, the higher the level of advancement of the people who live in those planes.

8.6.1 Planes of the Bottom Half of SD2

The Soul-Consciousness Stage is divided into four sub-stages, which occupy the bottom half of SD2. These sub-stages are merely listed here with a brief synopsis; a more detail described is provided in the next chapter.

1. The <u>Commencement Sub-stage</u> occupies the first quarter of the bottom half of SD2.

2. The <u>Multi-Sensory Sub-stage</u> occupies the second quarter of the bottom half of SD2.
3. The <u>Physical Zenith Sub-stage</u> occupies the third quarter of the bottom half of SD2.
4. The <u>Etheric Sub-stage</u> occupies the fourth quarter of the bottom half of SD2.

8.6.2 The Commencement Sub-stage

People that have transitioned from the Human Consciousness Stage to the Soul-Consciousness Stage occupy the beginning planes of the Commencement Sub-stage. Many people at the lower planes of this quarter live earth lives along side us at our present time. These planes are characterized by peace and tranquility since the people of these lower planes are utterly nonviolent. These people are generally well educated and professional. In earth lives they have professional jobs and in SD2 they continue the same general way of being.

8.6.3 Planes of Multi-Sensory Sub-stage

These people have developed their senses so that they are perceptive of things not only in the Physical Domain but also in the Astral Domain. These people are very advanced intellectually and technically. They have discovered most of the physics, chemistry, and biology of both the Physical Domain as well as the Astral Domain. Their pastimes in SD2 are, for the most part, centered on intellectual pursuits. These planes are characterized by peace, harmony, beauty, intellectual interest and the love of God.

8.6.4 Planes of Physical Zenith Sub-stage

People of this sub-stage are the highest level of humans that incarnates into physical bodies to live earth lives. They are of the highest intellectual developments that can still benefits from living earth lives, and they are masters of all physical and astral comprehension. They are masters of thought dynamics in the astral world and can read astral energy patterns and thoughts. Their senses have developed to the highest degree of perception commensurate with the needs of their environment. This sub-stage of Consciousness has developed a true paradise on earth as well as in SD2. They have also developed spiritually to a high degree where the love of God is paramount in their being. To us now, these people would appear as superhuman if we were to encounter them in person. This is so because their powers are beyond our comprehension.

8.6.5 Planes of Etheric Sub-stage

These people do not live earth lives anymore because earth life can no longer offer any more experiences that can contribute to their ongoing development. All of their growth from now on will be done in Spiritual Domains. The planes of the fourth quarter are flooded with abundant love, beauty, harmony, and peace. There is no such thing as discord, hostility, or conflict of any kind. People of this sub-stage have completely conquered the negative animalistic emotions, phobias and bad habits that plague us at our level of development.

There are exceptions to the rule of living their entire lives in SD2, there are instances in which a few of these people incarnate and live on earth. They do not come to earth for notoriety or fame and fortune, instead they come to help us get through some dilemma or hard times on earth. They usually keep a low profile, and remain almost hidden from the public at large.

8.6.6 Planes of the Top Half of Spiritual Domain 2

People of Cosmic Consciousness Stage live in the top half of SD2, and they no longer have to live earth-lives. There are a few exceptions where every now and than one or more of these advanced Consciousnesses incarnates into a physical body. The founders of the dominant religions of earth today such as Jesus, Mohammad, and Buddha, came to teach us from these higher realms of SD2. Some of the prophets of the Old Testament also came to us from these realms. Many Consciousnesses, which we call Angels, Archangels, Spiritual Teachers, and Spiritual Guides, reside in these higher planes.

8.7 Characteristics of Life in the Top Half of SD2

The top half of SD2 can be characterized as BLISSFULL. Relative to our way of thinking, SD2 is truly Heaven; it is truly Spiritual Enlightenment. The higher up you go within the planes of SD2 the more wondrous it becomes; superlatives that can be used to try to describe these wonders are beauty, splendor, serenity, sublimity, tranquility, glory, joy, harmony, peace, purity, radiance, rapture, epiphany, and more that we can not even imagine.

People of this stage of development have conquered the old negative animalistic emotions, phobias, and bad habits and are no longer handicapped by them. Consequently their environment is clean of negativity. Only the positive end of the emotions exists at this level. Remember that every emotion operates on a sliding scale from negative to positive. At the top levels of SD2 all the emotions have reached the positive end of the scale so that only virtuous characteristics exist. Only the higher levels of the virtues such as love, beauty, joy, peace, kindness, compassion, honesty, and harmony are present here. There are many more virtuous features that exist here but they are beyond our comprehension and it will be a long time before we can even begin to understand them. The Etheric landscape of SD2 is well on its way to absoluteness and is leaving relativeness behind.

8.7.1 Love in the Top Half of Spiritual Domain 2

The experience of love in SD2 is thousands of times more blissful than is possible in earth-life. We are social beings and instinctively long for true companionship. True companionships with strong bonds of love are the norm. Love is not sex and there is no such thing as sex in SD2 but there are emotions of euphoria and ecstasy that resemble our idea of sex on earth. The difference is that love is true love and the emotions of euphoria and ecstasy are much more wonderful than we could ever imagine here.

Evolution of Consciousness

Words in our limited vocabulary cannot express the loveliness of love in SD2. The highest imaginable ideals of love, joy, bliss, and happiness in earth-life is a vague perception of the true reality of life in SD2 and even more so in SD3. Life in SD2 is true Nirvana for all who live there, but the beauty of it is that we are all destined to eventually reach those heights.

8.7.2 The Creative Instinct

SD2 is a true paradise; it is utopia in Heaven, but we have to develop a great deal more before we can fully appreciate its significance. Mother Nature is never idle and neither are the Consciousnesses at all levels of SD2. The creative instinct of the Consciousness comes from Nature Herself; consequently Consciousness is always enticed into creating, doing, performing, becoming, making, and achieving, forever onward without ceasing, through eons of time. The end result of all this activity is the continued progression to the ultimate goal of "Evolution of Consciousness". The people of the Cosmic Consciousness Stage, as advanced as they are, are still developing and growing in their environment. They always have serious work to do and they do it with pleasure. The instinctive creative impulse is given every opportunity to put the old ideas to work and to make them bloom and bear fruit. This is where ideas are made real. There is no room for idleness.

8.7.3 Mental Imaging

The reality is that all landscapes in all planes of SD2 are mental images created by the people who live there. The huge diversity of scenery that a person sees when traveling through the planes of SD2 is made of thought forms. People of like thoughts and aspirations tend to gravitate towards each other. When these people of similar thoughts come together, they tend to generate mental images of sceneries or landscapes that are commensurate with the average thought forms of the group. The sceneries are a composite of the typical thoughts of the people that live there. Groups of likeminded peoples create sub-planes, and there are many of these sub-planes in each plane. In earth-life the environment largely molds the person's personality, but in the SD2 the people that live there make their environment with their composite ideals and mental images. It stands to reason that the higher the plane the higher the development of its inhabitants, and consequently the higher the quality, beauty, and wonder of the landscape or panorama.

8.7.4 Dimensions

The higher the Consciousnesses are on the planes of the universe the less we know about them and consequently the less we can say about them. The reason for this is that their intellect advances out of the range of our experiences on earth and therefore out of range of our understanding. Even if these people could relate their knowledge to us, we would not have the vocabulary to express this knowledge. Their environment is multidimensional and we cannot cope with anything that has more than three spatial dimensions. We are three-dimensional beings, whereas they are multidimensional.

It must be remembered that these advanced people of SD2 are the same as us. The only difference is that they have already gone through the experiences that we are now going through. We will eventually get there, just as they have. We will eventually saver the wonders of SD2. Each step of the way is arduous, but each step also brings us immense satisfaction and joy. We will not falter because it is our destiny to continue to advance to these lofty levels of the universe.

8.8 Spiritual Domain 3

Spiritual Domain 3 (SD3) is the highest domain in the hierarchy or the universe, and is populated by peoples of the highest and most advanced Consciousnesses of the universe. These people belong to the Spiritual Consciousness Stage. There are millions of planes in SD3, and there are billions of people of the Spiritual Consciousness Stage who live in those planes. These people are reaching the peak of human evolution in the universe, and are close to

Evolution of Consciousness

completing their journey. They continue to work hard and grow and develop until the end of the journey.

All the wonderful things that we mentioned about SD2 also apply to the planes of SD3, except that they exist in even more majestic splendor and glory. For instance, beauty is more beautiful, love is lovelier, companionship is more wonderful, life is grander, and so is the case with every condition. Divine Love is the rule in SD3 and Heaven is more Heavenly. We do not have the words or metaphors to relate the wanders of SD3 because the difference in Consciousness is so great. Their level of Consciousness is beyond our capability to comprehend. A crude analogy is, "People of SD3 are as far above our level of comprehension us as we are above the level of comprehension of the insect."

The highest planes of SD3 are the highest planes of the universe. Some characteristics of these planes are that they contain the highest frequencies of the universe, they are approaching absoluteness, they have many dimensions, they are unimaginably glorious and they are the residence of the most advanced Consciousnesses in the universe. These Consciousnesses are on the verge of completing their evolutionary journey and their final step is to merger with the Spirit and Soul. After that they merge with God and leave our universe forever. There is no knowledge of what happens after that because they enter the unknowable Realm of God.

Chapter 9

Consciousness

The six stages of Evolution of Consciousness are:
1. Plant Consciousness Stage – No awareness
2. Animal Consciousness Stage – Awareness of the physical.
3. Human Consciousness Stage – Awareness of the Physical and mental.
4. Soul Consciousness Stage – Awareness of the physical, Mental and Spiritual.
5. Cosmic Consciousness Stage – Awareness of being a microcosm of the universe.
6. Spiritual Consciousness Stage – Awareness of being an integral part of the Divine.

9.1 Plant and Animal Stages of Consciousness

9.1.1 Plant Consciousness Stage

Consciousnesses get their start with the ultra primitive and embryonic plants. It is like a Divine generator that is continually creating new embryonic Consciousness at the very beginning of life. This creation can be visualized as a vaporous plume that is being produced and which animates the bodies of the simplest single celled plants and animals such as bacteria and gives them life. It is the beginning of life. At its earliest point, plant Consciousnesses has absolutely no sense of awareness whatsoever; they have zero awareness of being. Yet there is enough Consciousness in even the most primitive plants to get the process of life started. It is a process that takes millions of years to complete, but it is destined to culminate in wondrously high levels of being.

There are similarities of the lowest of plant Consciousness to even the highest stages of Consciousness. The similarity is that all Consciousness is made of the same basic materials, which are wavelets that vibrate at incredibly high frequencies, but after that the differences are enormous. Since there is no sense of awareness in plant Consciousness, about the only thing that it can do is to accumulate data of the experiences gained by the life of each plant. These experiences are minute but over time (thousands or even millions of earth years) they contribute to its gradual evolution. Refer to the illustration of the dewdrop analogy in Chapter 6, Section 6.7.4.

The Plant Group Consciousness is what gives the species of plant its place in life, and each species of plant has its own Group Consciousness. The Group Consciousness remains with a species of plant for many generations of plants and with each generation it gains something that is akin to experiences. These experiences causes the Plant Group Consciousness to grow and evolve, and eventually it learns all that there is to learn within that primitive species of plant. When that happens, the Plant Group Consciousness moves on to the next higher species of plant and goes through the same process all over again. Each time it reaches the highest point of knowledge gathering within a species it moves on the next higher species. This goes on until it reaches the highest point of the Plant Kingdom and then it is time to transition to the Animal Consciousness Stage.

Group Consciousnesses in plants have nothing to do with the guidance of plant life because, at this point, Consciousness is not smart enough to guide anything. Plant life is guided strictly by the PIOM. Plant Group Consciousnesses have their existence in the Physical and Astral Domains, and no existence in SD1. For example, when plants die they simply decompose into their original mineral content to be used again by other plants. The PIOM goes to temporary storage to be used immediately by some other plants of the same species. The Group Consciousness never dies; it continues its journey moving from species to higher species.

9.1.2 Animal Consciousness Stage

The Animal Consciousness Stage is a gigantic advancement over the Plant Consciousness Stage. Animals have a brain and nervous system which gives the animal mobility and they also have the AIOM which is more advanced than the plant's PIOM. Animals also have Group Consciousness for each species, and there are many levels or degrees of advancement in Animals Group Consciousnesses. They vary from the lowest forms of animal organisms to the advanced animals such as the higher mammals and even the great apes. The higher mammals referred to here include most of the domesticated mammals such as the horse, cow, pig, dog, cat, etc.; these are animals of relatively high intelligence. Some other high intelligence mammals, which may in some cases also be domesticated, are the elephant, dolphin, whale, primates (monkeys) etc. The highest levels of the mammals other than the human are the apes such as the gorilla, orangutan, and the chimpanzee. These animals have a relatively high degree of intelligence along with feelings, emotions, desires, personality and other attributes that belong to the human, but they are of a lower grade than that of the human. These animals, even though they have a high degree of intelligence, and know how to do many things, they are not smart enough to know that they know. They live mostly on an instinctive basis.

The higher levels of the Animal Consciousness Stage have awareness of the physical. The horse for instance may know that it is cold in some cold winter night, and he knows how to look for the warm barn or shelter where there is some protection from the cold. But, if the owner were to leave the horse tied to a tree out in the cold the horse would not be able to mentally analyze the situation. The horse would not know how to wonder when his owner will come to rescue him from the cold, or to think of how cruel it is to keep him out there. The horse would not know how to wonder if he will be left out in the cold again the next day, or how to be jealous of the horses in the warm barn. The horse is not able to pity himself or to wonder if such a life is worth living. The horse is aware of the physical pain and discomfort but is unaware of any mental pain.

The horse knows how to look for food and shelter, and how to protect itself from predators. In the wild the horse also knows how to create family social groups and how to organize hierarchical leadership societies, but all these are controlled by the AIOM and not the Consciousness. Many of the higher animas have skills that resemble the sophistication of the horse. These animals cannot think of themselves as knowing; they do not know that they know. Animals can experience the sensation of physical pain and discomfort of the outer world, but they cannot shift that to the inner state of being. Animals do not have the level of Consciousness to know themselves.

All animals are evolving from lower levels to higher levels within their ranks. The Animal Group Consciousness moves up within the species as explained previously for the Plant Group Consciousness. When an Animal Group Consciousness reaches the highest level possible within a species, it then jumps to the next higher animal species. In its process of growth, the Animal Group Consciousness keeps moving on from lower species to the next higher species over millions of years until it finally reaches the highest possible level on Animal Group Consciousness. When the animal reaches this highest possible point within the Animal Kingdom, it must move on to the next programmed level, which is the lowest level of the Human Consciousness Stage.

Animal Group Consciousnesses have their existence in the Physical and Astral Domains, and no existence in SD1. The reason for this is that Group Consciousnesses are always present in living animals only. In other words, when the animal dies, its Consciousness does not go to SD1, it merely merges with the rest of the Animal Group Consciousness. At the time of death of the animal, the physical body as well as the astral body immediately starts their normal mineral decomposition and dispersion into their physical and astral elements to be reused later by other living organisms. The AIOM of the animal goes to nature's storage, also to be used by other animals of the same species later. The AIOM may be updated by nature as required before entering the next animal for the duration of the next life.

9.2 Mechanics of Transition from Animal to Human

When the Animal Group Consciousness reaches its highest possible level of evolution it is ready to transition and to become a Human Consciousness. The highest level of Animal Group Consciousness could be the great apes or possibly a domesticated pet. Consciousnesses from the domesticated pets are prime candidates to transition because they have acquired some of the human traits after living with humans for extended durations. Generally what happens is that a section of the Animal Group Consciousness breaks away, under the direction and supervision of NGI, and becomes a Human Consciousness. After that the process requires a search for the right parents and the right environment. The New Human is guided by the "Law of Attractions" in search of parents and environment that are commensurate with its own primitive level of development. When the right parents and environment are found, the New Human Consciousness enters the body of a total, complete, and modern human baby. There is no intermediate step for the Consciousness; there is no missing link to go through before entering the human body. It is important to note that in animals the Group Consciousness moved up the ladder of evolution by jumping from one animal species to the next higher animal species, but in humans there is only one human species. Once the Consciousness reaches the level of the human, it evolves from sub-stage to sub-stage, but always within the same, one and only, human species. This one and only human body type fits all levels of human evolution on earth. Our single human species contains all the necessary equipment to accommodate the evolution of all levels of human Consciousness. It is a wonderful design.

As mentioned before, when the Consciousness first jumps to the human level, it has to look for parents and a society that is compatible with its earliest level of awareness. This means that the parents have to be primitive and have to be part of a primitive tribe that is just starting the journey of human evolution. At this early level of Consciousness these people are not much more advanced than the animal where they came from, yet paradoxically the difference between these "New Humans" and the animals is profound. The reason the change is so profound is that at the time of transition from animal to human the "New Human" gains eight human attributes that were not present in the animal. The eight attributes acquired at transition are as follows:

1. Soul and Spirit – These are absolute qualities of each human being; they are part of the Real Self, and are also eternal. Animals have Group Spirit and Soul.
2. Individual Consciousness – Plants and animals have Group Consciousness, humans acquire an Individual Consciousness, which is for eternity.
3. Intuitive Mind – This mind is available to primitive humans but not used extensively until the human advances to higher levels.
4. Intellectual Mind – This mind is used for thinking, reasoning, analyzing, and making decisions. The higher animals have a rudimentary Intellectual Mind but not sufficiently advanced to be classified as a true Intellectual Mind.
5. Complex Personality – Human personality is far more complex than the rudimentarily personality of animals.
6. Trainable Mind – The same basic AIOM of the animal also transitions to the human but it acquires the Trainable Mind, so that the HIOM is the summation of the AIOM plus the Trainable Mind. The Trainable Mind is what allows the human to learn to do the many thousands of repetitive things in a more or less automatic mode. Humans can thus learn such things as riding a bike, driving a car or using tools. The Trainable Mind also allows the human to learn speech.
7. Upright Posture – The upright posture of humans frees the hands, which allows us to do much more physical activities more efficiently. Free hands provide the humans with the

Evolution of Consciousness

agility to do many things and consequently the potential to learn and gain more experiences.

8. Large Brain – The new and large physical brain of the human is the tool that allows the Trainable Mind to do all the many more physical activities plus it also gives us the ability of speak, read, and write. The large brain and Trainable Mind combination also gives us the ability to use and manipulate the computer, which can be considered an important extension to our present intellectual resources.

These attributes are what makes the human enormously more advanced than the animal. The New Human has all the attributes of the human, and is a complete human from the very beginning. The only thing lacking in the New Human is the experience and knowledge that comes with time. Modern humans are much more advanced intellectually only because we have been around much longer. We have lived many more lives and consequently gained many more experiences, and as a result we are intellectually superior to the New Human, but you must not forget that each and every one of us were once at the same level of the New Human.

The New Human may be considered a savage by our standards, but it has all the necessary "know how" to make a good living in its new environment. At first the New Human lives mostly on instinct like the animals that it came from. Yet, it has all the instinctive skills to find food and shelter, to be a part of its family and tribal society, to find a mate, have a family, and raise a family. The New Human learned these instinctive skills while it was still in the animal stage. The New Human is generally a peaceful person and a good-natured person; he or she is not warlike or bloodthirsty, as Hollywood might tend to portray them. To begin with, New Humans are close to nature. They may be ignorant, but now they have *"all the tools necessary"* to advance up the ladder of evolution. They live a simple life, but with an important purpose, which is to do things, gain experiences, gain knowledge, gain wisdom, and advance. The New Human is very materialistic and tends to spend little time in SD1 after a physical death. The physical environment is what the New Human knows best and what he or she is most comfortable with, consequently soon after they transition to SD1 they are ready to come back. Earth life is where they thrive.

9.3 Human Consciousness Stage

The Human Consciousness Stage starts immediately after the Consciousness transitions from animal to human. This transition is enormously important because it is the start of each human individual on the evolutionary path. From its earliest beginnings, everything for the human is forward advancement, seemingly never ending, and the final outcome will be glorious. The adventure of the Evolution of Consciousness eventually culminates in the merger with God.

At transition the New Human is given all the necessary tools to develop and grow, but the New Human is not given knowledge, wisdom, or full awareness of being. The New Human has to arduously struggle to gained knowledge, wisdom, and awareness over eons of time. It is the job of the Human Soul and Spirit to guide the New Human Consciousness through its trials and tribulations of growth. In other words the Human starts out ignorant and has to gain wisdom on its own with the guidance of the Soul. The Soul is like a good parent who wants its child to grow up properly but knows that it cannot make the child learn if the child does not want to learn. The Consciousness, like a child, has to learn on its own, at its own pace, and the Soul, like the parent, can only provide guidance. The Soul cannot learn for the Consciousness.

It may be difficult for us understand why God did not start us off knowing everything. I suppose that He could have made us smart from the start since He can do anything He wants. But no, instead we have to start at the bottom, completely ignorant, and from there we have to learn everything. We have to work our way up one step at a time over thousands of years until we eventually reach the point where we know everything. Apparently that is the way of God.

9.3.1 Sub-Stages of the Human Consciousness Stage

The Human Consciousness Stage is the stage that we know best because it is our present stage of development. As a matter of fact, it is the only stage that we are really aware of. Our history books are based on the accounts of what we have done through our past, which in essence is the Evolution of our Consciousness thus far. It should also be noted that the Soul and Spirit, being of God's substance, are absolute entities and therefore are not hampered by time and space. In other words, what may seem to the Consciousness as eons of time to grow up is like the blink of an eye for the Human Soul and Spirit. This is one of the tricks of relativity that is difficult for us to understand at the present time. This book subdivides the stage into six sub-stages for the purpose of explanation and clarification. Each sub-stage is briefly discussed, not for the purpose of documenting human history, but for the reason of showing the step-by-step progression of the development of Human Consciousness.

- Sub-Stage 1 – **New Human**, equivalent to the human of the Stone Age or Caveman.
- Sub-Stage 2 – **Early Human**, equivalent to the human of the Bronze Age.
- Sub-Stage 3 – **Primitive Human**, equivalent to the human of the prehistoric civilizations of ancient Persia and ancient Egypt.
- Sub-Stage 4 – **Intermediate Human**, equivalent to the ancient classical Greek and Roman civilizations and extending through the Renaissance period of history.
- Sub-Stage 5 – **Civilized Human**, equivalent to the period after Columbus discovered the New World and extending through to the present time.
- Sub-Stage 6 – **Refined Human**, equivalent to the well educated 20th and 21st Century person.

The first three sub-stages are prehistoric and are known to us through the science of archeology, and the last three sub-stages are known to us through recorded history, which we can find in our history books.

9.3.2 Sub-stage 1 – New Human

The beginning levels of the "New Humans" are equivalent to the savage or barbarian, which are not much higher than the highest animals. The term savage may seem like a derogatory or bigoted term but we must remember that we all had to go through this sub-stage. It is true that it seems a contradiction when we say that the New Human is tremendously more advanced than the animal, and then we say that this New Human is not much smarter than the animal. The big difference is that the New Human has the tools necessary for further progress, but only lacks the human experience and knowledge that can only come with time. The lowest levels of humans live mostly by their instincts, and are guided by their lower instinctive emotions. New Humans are materialistically inclined and live mostly in physical earth planes having little to do with SD1. The physical body and physical environment is all there is from their point of view. At first they have no concept of God or Soul or Consciousness. These people perceive their physical world through their five senses of sight, smell, taste, sound, and touch. Their world is only what their five senses tell them and nothing more. At this early stage of Consciousness, humans are able to think of the "I" as I see, I hear, I feel, I taste, I want, etc.

The New Human is always born into a primitive tribe of humans where the parents are also recent arrivals to the human race. The main reason that the New Humans select primitive tribes, primitive society, and primitive parents to be born into is because of the demands of the "Law of Attractions". If they were to select anything else they would be incompatible, out of place, and unhappy.

Before Columbus discovered the New World there were vast numbers of primitive tribes that were suitable for New Humans. Soon after that, Europeans started claiming the lands of those primitive peoples and colonizing for the purpose of extracting wealth. During this early colonial era, the majority of these primitive peoples were wiped-out. Many were killed in battle, many died due to subjugation or imprisonment, and many more died due to European disease for which they had no immunity. In the 21st Century there are few places in the world where they can live in peace. In this modern age there are still some primitive tribes in the Amazon rainforests of Brazil, in some remote areas of Africa, in isolated mountain areas of New Guinea, and in a few other remote areas of the world.

At first New Humans are normally shy and seem somewhat out of place, as they try to fit in their new human environment. Starting from scratch, they are ignorant of their new place in life and have a lot to learn, but their eagerness to learn at this stage is strong. New Humans are social beings and united to their family and tribal society. They have no trouble adjusting to tribal social order and political organization. This is so because many of the social skills had been learned while still in the higher levels of the Animal Consciousness Stage. For this same reason, New Humans are capable of having their own families and providing for their needs and well-being. They are capable of providing the essentials such as food and shelter as well as love and guidance for their family. To the New Human, speech is a new tool that is somewhat awkward at first and requires some special attention, but once mastered it becomes indispensable. The larger human brain is what makes human speech possible because speech is physically intensive. Speech allows Communication of complex thoughts, which are necessary for human development.

The primitive animalistic emotions such as fear, hate, anger, etc. are very acute in the New Human because they are direct inheritance from the animal. They still need those animalistic emotions and feelings for their survival. New Humans are not vicious, or bloodthirsty and cruel; they kill only for food even though, in some cases they may kill during tribal squabbles. They do not abuse their ability to kill. Human barbarism comes later in more advanced levels of human evolution. However, cruelty or aggressive behavior is not an offense for the New Human; not anymore than it is an offense for a tiger that killing its prey for food. The New Human, because of his ignorance, is not responsible for wicked actions, but the more advanced human does become responsible for wrongful actions because he should know better. The Law Cause and Effect comes down harder on advanced people for wrongdoing than it does on primitive people.

Generally New Humans are born and live their entire lives in the same primitive villages. They seldom venture out more than 50-mile from their village during their entire live. Migrating to distant places does occur, but it is not the norm. New Humans normally live in peace. Most of their lives are spent gathering food, hunting for food, and socializing with their family and relatives within their village. If left alone, they are happy people for the most part. Contrary to the movies and popular beliefs, New Humans are normally not cannibalistic or cruel, and they do not participate in gory human sacrifices. New Humans do occasionally participate in tribal wars or family conflict, but seldom are there human deaths as a result. If people are killed during a conflict, they are few in number. Primitive people do not have the military weapons, techniques, tactics, or desire to kill large numbers of people. Modern people, the so-called civilized people, do have the means, and do commit large-scale human atrocities.

9.3.3 Sub-stage 2 – Early Human

The Early Human is equivalent to what our history books and archaeology would classify as the hunter-gatherer and their period existed through the Bronze Age. Many of the characteristics that pertain to the New Human also apply to the Early Human with some notable advances such as greater use of tools, primitive farming and domestication of animals. The Early Humans, after having lived through many lifetimes on earth, have become masters of their environment. If there is no outside interference with their lives, they live happy and tranquil lives. They have great respect towards nature and take only what they need without ever abusing, destroying, or polluting nature.

The Early Human lives on instinct to a great extent, but the intellect and reasoning power is already making big strides forward. They are close to Nature but they do not think in terms of ownership of land. These people have more and better tools for primitive farming and hunting. Some tribes are nomadic whereas others basically stay in one place. The more advanced civilizations of this sub-stage use copper and bronze tools. Tools are used for hunting and farming, but occasionally they are used as weapons of war.

Religion is a human instinct and primitive versions of religion flourish among the Early Humans. They normally worship many gods such as the Sun God, Moon God, Rain God, Wind God, Fire God, etc. They may also have symbols to worship, such as a stone or wooden idles.

9.3.4 Sub-stage 3 – Primitive Human

Primitive humans are equivalent in advancement to people of the late prehistoric times, which include early biblical times, and the early civilizations of ancient Persia and Egypt. Writing and mathematics were in their infancy but well underway. Their livelihood was very diverse; they made their living in farming, merchandising, trading, fishing and many other trades. They saw the beginning of the early city-states with efficient governing organizations. The Aztec, Mayan, and Inca civilizations of the Americas are examples of this sub-stage.

Primitive humans were developing more sophisticated tools and weapons, but they were not a war like peoples yet. There was limited violence among primitive people, but they should not be considered violent or warlike people. They still lived close to nature and honored her. Most religions of the times consisted of many Gods, and many superstitions revolved around there concerned with illness and death. The early Jews and early Hindus got their beginnings around this period of time and of course they believed in only one God as they do now.

9.3.5 Sub-stage 4 –Intermediate Human

The Intermediate Human is equivalent in advancement to the classical Greek and Roman Civilizations and goes through the early Renaissance period in Europe. Of course, the Far Eastern civilizations also had equivalences to this sub-stage of evolution. The Intermediate Human has a well-developed written language and has the beginnings of questioning what we are all about and what our universe is all about. Science, philosophy, theology, ethics, art, music, etc. were starting to flourish. Social, political, and governmental organization also advanced significantly to the point where large city-states flourished and were common.

The Intermediate Humans with their growing intellect and awareness of self and the environment opened the path to greater development. The greater intellectual powers are beneficial but can also be detrimental to others as well as to themselves. The higher intellect gives the humans the ability to become masters of their environment, but also gives them the ability to become a greater beast than the beasts themselves. In Europe the "Intermediate Human" advanced the art of military weapons, techniques, and tactics to a large extent. This sub-stage starts to use their superior intellect to create not only great works of art, but also advance the ability to create excessive violence, cruelty, and savagery. The sheer brutality of the gladiators in the arena for the purpose of pleasure during the glory days of the Roman Empire is an example of this savagery. Even the most brutal animals were never as brutal as many humans of this sub-stage. Another example of these is the cruelty that war conquers would impose on the vanquished.

The lust for material goods and power become prevalent and as a consequence grief, suffering, and unhappiness also become prevalent for this sub-stage. The Intermediate Human saw the beginnings of the use of the intellect rather than instinct to conduct their lives. The dominant religions of the world today – Buddhism, Islam, and Christianity – got their start during this time.

9.3.6 Sub-Stage 5 – Civilized Human

The Civilized Human Sub-stage is characterized by the lust for wealth. This sub-stage got its beginning around the time of Christopher Columbus' discovery of the New World. That is when the Europeans started the exploration and conquest of the underdeveloped world. That saw the colonization and demise of the Aztecs, Mayans, and Incas in the Americas. Soon after that there was European colonization of Africa, much of the Far East and the subcontinent of Asia. Many of the primitive races of the world at that time were wiped out within a relatively short period of time. The educated, affluent, and aristocratic people of Europe best represent this sub-stage in the early days.

This sub-stage is also noted for its great accomplishments in almost every phase of human endeavor. These include huge advances in science, philosophy, theology, economics, government, art, theater, music, technology, transportation, medicine, and many others. It also includes enormous increases in wealth and pride in nationalism. The advances in trade, industrialization, business, and affluence contributed significantly to the rapid growth of Consciousness. In the 20th century, the majority of the people of the world could be classified as belonging to the Civilized Human Sub-stage.

The Civilized Human, while it is accompanied with great deeds of goodness, benevolence, and righteousness, is also accompanied with great evil, violence, and atrocity. The main reason for this is that their intellectual capacity is advanced but their control of the animalistic emotions and feelings are not under control. As a result, great pain, suffering, agony, misery, and poverty are caused in widespread areas of the earth. Weapons of war and destruction become sophisticated and abundant. The two most devastating wars of human history, World Wars 1 and 2 were fought during the 20th Century. This devastation was the result of the huge advancements in the technology of war, war strategies, and the accompanying advancement in Human Consciousness. Greed and the overwhelming desire for material wealth and power are common. The formation of the powerful business aristocracy, corporate conglomerates and organized banking enterprises tended to create great wealth for many. This also tends to cause a climate of violence for some, and a climate of poverty for others.

The overriding advancement of the Civilized Human is living by the wit of their Intellectual Mind rather than the wit of their instincts. This is a departure from the Primitive Human who lived primarily on instinct. The experiences of mental stress are a consequence of the newfound mental powers. Humans at this stage can come up with countless activities to cause distress for themselves, their family, and others. The huge advances in technology have accelerated the pace of the Evolution of Consciousness as well. As humans advance they build better technologies, and better technologies helps advance the development of Consciousness. This feedback loop feeds on itself and causes the pace of development to accelerate.

9.3.7 Sub-stage 6 – Refined Human

The "Refined Human Sub-stage" is equivalent to the affluent and educated peoples in the developed countries of the world in the 20th and 21st Centuries. At this sub-stage people still have great desire for material wealth and power, but no longer at the expense of violence and destruction. The desire for peace, happiness, joy, beauty, knowledge, and wisdom are now within their grasp. The Refined Human now lives almost entirely on the direction of the Intellectual Mind and its reasoning abilities. The lower animal emotions are still prevalent but better controlled and suppressed, even though they do occasionally show their ugly ways. The human at this sub-stage is no longer a slave of the instinct. Greater intelligence also brings with it greater responsibility. According to the "Law of Cause and Effect", wrongful doing by an advanced person can cause greater damage and so consequently the punishment for wrongful doing is also greater. Because of the obligation toward the Law of Cause and Effect the Refined Human has to become more responsible for his or her actions.

The Refined Human now has a greater desire for self-improvement and personal advancement. People at this sub-stage of development tend to try to get more formal education. These people have a strong desire or yearning to search for clarification of the true self. They tend to read and study more books in self-help, in religion, in philosophy and other new age teachings. The hunger for more knowledge of Spirituality and the love of God is now more acute. The advanced persons of the Refined Human Sub-stage now gradually but surely start thinking of themselves as Spiritual Beings, and not as mere mortal physical bodies. After a long and arduous trip through all the sub-stages of growth that have taken many lifetimes and thousands of years, the journey is now almost over.

The next step is to make the transition to the next stage – the Soul Consciousness Stage. This transition is as profound as the transition made thousands of years ago from the Animal Consciousness Stage to the Human Consciousness Stage. This will be the beginning of a new adventure that will reach to new and more wonderful heights of peace, joy, love, happiness, and self-fulfillment. From here on there will be less violence and pain, and once started on to the new stage there will be no looking back.

9.3.8 Estimated Distribution of Sub-stages in our World Today

Our world today consists mostly of people of the fourth and fifth sub-stages. There are a few people that belong to the first three sub-stages and a few belong to the Soul Consciousness Stage. The estimated percentages of people of the various sub-stages in our world today are:

- Sub-stages 1, 2, and 3, make up about 5% of the present population of our world.
- Sub-stage 4, the Intermediate Human makes up about 20%.
- Sub-stage 5, the Civilized Human makes up about 55%.
- Sub-stage 6, the Refined Human makes up about 15%.
- About 5% of the population belongs to the lowest levels of Soul Consciousness Stage.

The reason that so few of sub-stages 1, 2, and 3 exists in our world today is because most of them have been wiped out by the more advanced Civilized Human Sub-stage. Our present civilized society is the most violent society in the universe. The bottom three sub-stages are basically nonviolent; generally they are violent only in defense and for their survival. Generally they only kill to eat. The Refined Human has basically outgrown violence and is unlikely to resort to violence except in dire emergencies. The more advanced Soul Consciousnesses Stage has completely eradicated violence; its environment is much more peaceful and tranquil.

9.4 Soul Consciousness Stage

Soul Consciousness Stage is the next higher stage of Consciousness and the transition from one to the other is almost imperceptible to an outside observer. The distinguishing factor of the people of the Soul-Consciousness Stage is that they now think of themselves as Spiritual Beings. They think of themselves as a "True Self" which exists within. They now know that they have a body and mind, but the True Self within is the Consciousness, Soul, and Spirit. At this level the person does not just hope or believe to be a Consciousness; it now knows positively and without a doubt that he or she is a Spiritual Being. This is not a mundane matter; it is profound matter. The transition to the Soul Consciousness Stage is as significant as the previous transition from the Animal Consciousness Stage to the Human Consciousness Stage. It is important to realize that by changing the way a person thinks it changes the way the person is.

It may seem contradictory when it is said that the transition to Soul Consciousness Stage is profound and then at the same time it is said that the transition is imperceptible. It is a paradox only because the transition usually takes place over a long period of time. Often the transition takes place over two or three lifetimes, but for people who are ready it can take place over a period of twenty to forty years of the present lifetime. In some cases it may be quicker than that. Sometimes it is difficult to move on because of past religious dogmas or social or cultural biases that tend to slow down the process. Most people feel uneasy during the time of transition because they feel as if they are abandoning their past, their traditions, and often they feel they are abandoning their friends and relatives who do not seem to understand. But, in due time the person feels certain and more comfortable about the new convictions and then proceeds on the new path. Once on the path to the new stage of being there

Evolution of Consciousness

is no turning back. There is a feeling of joy and exaltation that it is the right thing to do, and a feeling that the friends and relatives will follow in their own due time. It should be noted that for every transition period there is always a period of blending in from one stage to the next. At the present time in our history there are many people of the "Refined Human Sub-stage" who are reaching the highest level of the Human Consciousness Stage and will soon be getting ready to transition.

9.4.1 Sub-Stages of the Soul Consciousness Stage

The Soul Consciousness Stage like the Human Consciousness Stage can be divided into sub-stages for the purpose of illustration. Our knowledge of the Soul Consciousness Stage is not as extensive as what we know about the Human Consciousness Stage, but we know the main principles to give us a good idea of what to expect. For the purpose of this book, the Soul Consciousness Stage is divided into four sub-stages as follows:
Sub-Stage 1 – Commencement Sub-Stage
Sub-Stage 2 – Multi-Sensory Sub-Stage
Sub-Stage 3 – Physical Zenith Sub-Stage
Sub-Stage 4 – Etheric Sub-Stage

The first three sub-stages are associated with living physical lives on earth – their life cycles between earth and SD2 – and those people of fourth sub-stage live in SD2 only. This is generally true but of course there are always some exceptions. Consciousnesses at this level have a choice, they may choose to live earth lives or they choose to live in SD2 exclusively, and there are many among sub-stages 1, 2, and 3 who do not wish to reincarnate anymore. These people live their entire lives as ethereal bodies only.

9.4.2 Sub-Stage 1 – Commencement Sub-Stage

At first glance, the Commencement Sub-stage is almost indistinguishable from the Refined Human Sub-stage of the Human Consciousness Stage, but underneath it all there is a huge difference. The main difference being that the person is now conscious of being a Spiritual Being. They now realize that the body and mind are important entities that are to be cared for and glorified but they are temporary; the Consciousness is the Real Self and it is eternal. The realization is just now sinking in that the Consciousness is the most important entity in the universe and that the universe was created exclusively for the purpose of providing a platform for Consciousnesses to evolve in.

Some of the characteristics of the Commencement Sub-Stage are that these people are now more virtuous. There are some in our society who consider the words, such as virtuous and righteous, to be a derogatory depiction of someone, well that will have to change before progress can be made, like it or not. At this sub-stage the person becomes more aware of the plight of the downtrodden such as the sick, the poor, the illiterate, the elderly, and there is a desire to help. Kindness and compassion for all living things becomes second nature. For instance, the thought of killing a deer or a rabbit or anything else for sport is unthinkable at this stage. Voluntarism and public service becomes the norm; philanthropic pursuits to help the needy are also common among these people. Participation to help advance education and personal growth are also commonplace. Clean, productive, and compassionate living is the standard style for this early phase of the Soul Consciousness Stage. These people are also high-achievers; they give it their best in everything they do. For instance, they are hard workers and are tops in their profession, as well as participating in civic activities, as well as sports, in music, in the arts, and many other high minded ventures. These people have great amounts of self-discipline. They are well under way to conquering negative emotions, bad habits, and phobias; they also maintain good and healthy eating habits.

On the religious level these people worship the one God who is Omnipotent, Omniscient, Omnipresent and Absolute. The love of God and everything Divine becomes paramount in their lives. There is a newfound awareness of being a part of the Divine, and that everything of Nature is also a part of the Divine.

The process of living consciously becomes the rule. What this means is that there is less living per the dictates of the Instinctive Mind and more the dictates of the Consciousness. The Consciousness finally becomes the master of the mind and body. At this stage the Consciousness dictates to the Intellectual Mind what to do, the Intellectual Mind dictates to the instinctive mind (HIOM) and the HIOM has to obey. By this time the lower negative animalistic emotions are for the most part under control. It is not likely to find a person of this level having temper tantrums, or making character assassinations, or defrauding other people. People of the Commencement Sub-stage are exemplary to the people of the Human Consciousness Stage. At this sub-stage, the person knows that the Consciousness lives within when living an earth life and the physical body may be called the "Temple of the Consciousness". During an earth lifetime these people are aware that they live in a relative world in which there is time and space.

People of the Commencement Sub-stage can coexist with all sub-stages of the Human Consciousness Stage, but tend to have some trouble with the Intermediate Human and the Civilized Human sub-stages. The reason for this is that they tend to be a little rough around the edges and have trouble understanding the person of the Soul Consciousness Stage. Many may tend to be of the "Red Neck" verity and have choice words like "Mr. Goody, goody" or "Mr. Righteous" or other words to show sarcasm. The truth is that where people of the Commencement Sub-stage become numerous or where they dominate, peace will ensue. Wherever this people rule, such things as crime, corruption and sinfulness diminishes. Whenever these people govern the world, the world will become free of war and violence.

9.4.3 Sub-Stage 2 – Multi-Sensory Sub-stage

When the Consciousness reaches the "Multi-Sensory Sub-stage" the brain undergoes a slight physical modification in the pineal gland and the Human Instinctive Operating Mind (HIOM) undergoes a mental modification as well. These modifications are designed to increase the sensory capability to perceive the Astral Domain. With this modification they are able to see, hear, feel, taste, and smell the Astral Domain. Mother Nature provides this for the purpose of improving the potential for gaining new experiences; knowledge, wisdom, and awareness while living earth lives. These modifications are relatively slight but in actuality they provide this level of person with an enormous advantage. People of the Multi-Sensory Sub-Stage would seem as super humans to us, if we could encounter them in real life. There perceptive abilities far exceed our capacity to perceive and understand things.

Actually the new senses consist of an enhancement of the existing five senses. This modification will allow the person to see equally well in the Astral Domain as well as in the Physical Domain. This enhancement of the senses is equivalent to opening up another new universe in which to live in. Opening up the Astral Domain to our already huge living environment will just double our potentiality for new experiences and growth. In the Astral Domain the person will be able to see much farther in all kinds of weather and also hear a wider range of frequencies from the very low to the ultra high frequencies over a much longer distance.

The ability to see in the Astral Domain also gives the person the ability to see human auras. Human auras are the result of human "vital energy" (also called astral energy), which is made of frequencies in the astral spectrum. The human aura is composed of many auras since each component of the human produces its own aura. For instance, the physical body has an aura, the astral body has an aura, and the mental has an aura; all these auras coexist and overlap around the physical body at all times. A person that can see auras can see them in various colors and can learn to interpret the physical health, fitness, and also the mental or emotional disposition of another person. This is so because certain astral colors and brightness of colors represent certain conditions of the body.

Another important benefit gained by the Multi-Sensory human is the ability for mental communications. The modification of the pineal gland is what makes this possible. This change allows the brain to become more receptive to thought communication, which are called Extra Sensory Perception (ESP). ESP is actually possible at our level of development, but most of us cannot receive and interpret thoughts efficiently enough to be of significance. Physic powers such as clairvoyance and mental telephony are widespread among the Multi-Sensory people. They communicate with each other face to face or over long distance anywhere in the world, and communications is also common with those in the spiritual world. Wireless phones will become obsolete. These people can also do "out of body travel" whenever they wish. Consequently our modes of transportation such as cars and airplanes also become obsolete.

This sub-stage can coexist with people of the Human Consciousness Stage, but it is not an easy coexistence because the differences are too great and the friction due to misunderstanding on the part of the lower levels of Consciousness causes too much grief. Multi-Sensory people learn reverence and compassion for all life. To do harm to any other form of life, be it human or animal, becomes unnatural, and it becomes unlikely for this person to participate in evil deeds. At this stage the person is not as likely to gain bad cause and effect since most of the causes and effects from now on are good. Remember that good and bad are relative terms, which in this case good means better than it is for us in the Human Consciousness Stage. Evolutionary progress accelerates to a large extent at the Soul-Consciousness Stage.

9.4.4 Sub-Stage 3 – Physical Zenith Sub-Stage

The people of the Physical Zenith Sub-stage" are the highest level of human that still incarnates into physical bodies, but they live earth lives only if there are practical experiences to be gained. If not, they can live out the rest of their development in SD2 only. The benefits of living an earth life may be in paying off some existing Karma, or gaining some experiences that have not been realized yet, or completing some adventure that was left unfinished with some companion in the past. There could be any number of reasons.

The Physical Zenith Sub-Stage creates what may be called a paradise on earth. The rule is peace, tranquility, happiness, beauty and love for all. At this level of Consciousness there is no violence, no wars, no terrorism, no fraud, and no animosity of any kind. Living with these people is like living in a peaceful and beautiful paradise. However, their idea of paradise is very different from our idea of paradise. In their paradise they do not live in huge mansions and they do not have fast cars and glamorous women tanning by the swimming pool. Instead they live in commune with Mother Nature. They do not pollute and they do not overpopulate the earth. They do not own the land instead they revere the land, the flora and the wild life. To them Mother Nature is of Divine Origin and should be revered as such.

These people do not coexist well with people of Human Consciousness Stage because the gap in the level of Consciousness is too great. The problem for us is that they are too good and we cannot stand it. For instance, we could not stand for those folks to know all the foul thoughts that run through our heads at all times.

These people can communicate verbally but for the most part they use mental communicate. In this society there is no place to hide because all thoughts are public domain. Everyone immediately knows all thoughts and all emotions, and as a result of these there is no ill will, or hate, or fraudulent activity. If ever there is contemptuous thought, it is exposed immediately and properly dealt with. These people have conquered the old animalistic negative emotions consequently there are no animosities. Only positive thinking, positive emotions, and positive activities are present here.

People of Physical Zenith Sub-Stage are conscious of being a Consciousness. They are conscious of past lives and can now recall the memory of past lives at will. They are also fully conscious of all activities of rebirth. For instance, they select their future parents consciously. They are aware of living in both the Physical and Astral Domains simultaneously and they are also fully conscious of their lives in SD2.

These people are intimately knowledgeable of the physics, chemistry, and biology of the Physical Domain plus they are also equally intimate with the physics, chemistry, and biology of the Astral Domain. At their level they have discovered the secrets of the "Unified Field Theory" or "Theory of Everything" which our scientists have been struggling with for years. The reason our scientists have been unable to complete the concept of the Unified Field Theory is because they lack the knowledge of the physics of the Astral Domain.

9.4.5 Sub-Stage 4 – Etheric Sub-Stage

People of the Etheric Sub-stage are the most advanced of the Soul Consciousness Stage and no longer incarnate into physical bodies. The main reason for this is that they are too advanced to get any real benefit out of living earth lives. Instead they live their entire lives in SD2; this is the fourth quarter of the bottom half of SD2. People of the Multi-Sensory Sub-stage and the Physical Zenith Sub-stage also have the choice of living their entire lives in SD2 and many do so.

Life in SD2 for the Etheric Sub-stage is filled with beauty and bliss. All the good things mentioned for life in SD1 and SD2 apply here. As usual however, as advanced as they are and as wonderful as their lives are, they still have to work hard to continue their evolutionary process. They still have some "veils of ignorance" that have to be exposed and converted to "veils of knowledge". This Etheric Sub-stage has for the most part conquered all the animalistic emotion. Each emotion has its own scale from negative to positive, and most of the emotions for each Consciousness are now reaching the maximum positive end of the scale. These people are no longer hindered by negativity. All phobias that were gained during the Human Consciousness Stage have also been conquered. So therefore in essence they have conquered all fear; that is, they have conquered the fear of fire, of heights, of snakes, of crime and all other possible fears that beleaguer our lives today. They have also conquered all bad habits that were developed in earth lives.

As mentioned these people now live their entire lives in SD2. Yet, occasionally some of these people incarnate to live among people of the Human Consciousness Stage to help in some dire circumstance. A few of these people may incarnate to help out during some natural disaster, or a terrible war, or pestilence, or political upheaval. Normally they do not come to seek fame and fortune, but because of their extraordinary deeds they often become famous. A couple of contemporary people that may fit in this category are Mother Teresa, and Mahatma Gandhi.

Existence of the Etheric Sub-stage is a strictly mental existence in SD2, but it is more real than the physical existence that we are familiar with. Their knowledge and awareness level far exceeds ours. Sometimes some of these people make themselves accessible to us from their ethereal existence through earthly mediums or channels and when this happens it is good. The reason it is so good is that these people are very advanced and can impart some of their superior knowledge. They can give us new information as long as it does not exceed our level of development.

There are many mediums in earth life that can readily contact people living in SD1 but that is not necessarily beneficial since these people of SD1 are not any smarter than we are. They are at the same level as us and consequently there is not much to be gained there. It is not recommended that people of SD1 be contacted; they have their own priorities of living their lives where they are and do not need any disturbances from us.

9.5 Other Worlds

It should be obvious from the preceding discussions that not all sub-stages are compatible with each other. The most visibly incompatible of all sub-stages are the Intermediate Human Sub-stage and the Civilized Human Sub-stages. Of all humanity, these two sub-stages are the most arrogant, the most vicious, the cruelest, and the most violent. They basically do not get along with anybody else. Of course, not all people are that bad, but there are enough of them to cause the overall rating to be awfully bad. It just so happens that these two sub-stages are the most dominant and the most abundant on our planet today. This is the reason why the previous three centuries in our recent history have seen the most horrific violence. World Wars 1 and World War 2, of the last century were especially horrific. Now, early in the 21st Century we started out with a bang when terrorists downed the Twin Towers in New York City. But as bad as it may seem to us now, it is actually normal for these sub-stages to be violent. In the larger scheme of things, that is typical for this level of development. This situation in part is caused because the Intellectual Mind is relatively advanced yet the overall ignorance level of the people is still very high. People at this level are intellectually advanced but emotionally inferior. Many of their actions are still influenced by the dictates of the subconscious animalistic emotions and feelings.

Before Christopher Columbus discovered the new world, there were many primitive tribes that populated our planet. There were tribes of every sub-stage from the New Human sub-stage to the Intermediate Human sub-stage. The reason this was possible in the world at the time before Columbus was that mobility of the people was still limited, and therefore primitive tribes in remote areas of the world could live in isolation and were safe. As the technology of transportation and weapons of war advanced the primitive tribes were doomed. Within 300 years after the discovery of the New World the population of the primitive tribes decreased by at least 85%. Now, in the 21st Century only a few primitive tribes can be found in the remotest areas of the world. Today there are a few primitive tribes of people that can be found living in remote parts of Africa, in the rain forests of South America, and in the mountains of New Guinea.

Because the two dominant sub-stages (the Intermediate and the Civilized Human Sub-stages) on earth today cannot get along with other humans they have to be isolated. The people of the bottom three sub-stages are primitive and consequently they are vulnerable since they are easy to conquer and destroy. On the other hand, the more advanced people of the Soul Consciousness Stage are humiliated by our arrogance and they also stay away.

So you may ask, "If so many people can not live in this earth with us, where do they live?" They cannot be denied there right to exist. The answer is that they live in "*other worlds*" that are similar to our world. The principle is that humans are constantly springing up from the bottom and moving up the evolutionary path. That is, the transitioning process from animal to human is continuous. The process goes on constantly, it has been going on for many thousands of years, and it never stops.

In actuality there are many worlds similar to ours in the universe. There are many worlds in this universe that sustain life that is similar to ours. These worlds are all at different stages of Evolution of Consciousness. Some worlds may have plants only, others may have plants and animals, and others may have plants, animals and humans at various levels of development. Some worlds have the New Human Sub-stage only whereas other worlds may have the New Human, the Early Human, and the Primitive Human Sub-stages. Several other worlds are like ours where only Intermediate Human and Civilized Human Sub-stages live. Yet other worlds exist where only the people of the more advanced Soul Consciousness Stage live. All these worlds are necessary so that Consciousnesses at all different levels can have the opportunity to exist and develop in their own isolation. The separation is made in order of compatibility of human sub-stages.

The separation is basically automatic and subconscious. It is conducted under the forces of the "Law of Attractions" and under the guidance of the "Nature's Guiding Intelligence" (NGI). All these worlds are similar to each other in many ways and yet there are many differences also. Some of similarities and differences between worlds are as follows:

- All worlds that contain life are of similar size, and have a similar sun at the proper distance to supply the energy needed to sustain life.
- They all have an atmosphere like ours with oxygen, carbon dioxide, nitrogen, hydrogen, water, and the like.
- They all have similar geographic features such as oceans, continents, mountains, lakes, forests, deserts and the like, but they are all completely different shapes.
- All worlds have life with a similar evolutionary history, but they are all at different stages.
- The generic evolutionary history in all worlds is similar but the details of the histories are very different.

Evolution of Consciousness

- The worlds that have plants and animals have equivalent Group Consciousnesses for plants and Group Consciousnesses for animals. The PIOM and AIOM are basically the same in all worlds with the exception of having slightly different versions. All worlds have plant and animal species that are similarly equivalent and can utilize the similar versions of PIOM and AIOM. Plants and animals look similar in all worlds but are actually of different species.
- Humans in all worlds are similar, that is, all humans are of similar size and stature, and all humans breathe air, drink water, and eat food. The difference comes in the coloration of the skin, the language and customs of the people in different worlds. Human history in all worlds is similar in general but very different in detail.
- Humans in all worlds are at different levels of evolution. Some worlds, for instance, only have humans of Sub-stage 1 of the Human Consciousness Stage. Other worlds may have humans of sub-stage 1 and sub-stage 2. Other worlds have humans that belong to sub-stage 1, sub-stage 2, and sub-stage 3.
- There are some worlds that are populated primarily by people of sub-stage 4 and sub-stage 5, and one of those worlds happens to be our world. Any world that happens to have predominantly people with this combination of sub-stages will, by definition, be a violent world. The reason is that these people are highly intelligent but not mature enough to know how to cope with their animalistic emotional makeup. People of sub-stages 4 and 5 do not coexist well with people of other sub-stages.
- Some worlds only have people of the Soul Consciousness Stage, and all of these sub-stages can coexist with one another. People of each of the three sub-stages will tend to segregate themselves into separate groups, but they can also coexist in the same locality. All worlds that have predominantly people of the Soul Consciousness

Stage are peaceful world. These worlds are paradise. The "Law of Attractions" has a big hand in determining where the people at different levels go to live.
- People of the New Humans Sub-stage and the Early Human Sub-stage can live peacefully and in harmony with people of the advanced Soul Consciousness Stage. The advanced people will treat the primitive humans with respect and will not take advantage of them.

These worlds are distributed throughout our universe in many different galaxies; there are immense distances from each other in physical terms. But, since there is no such thing as time and space in absolute reality, a person is just as likely to be born in one world as another and never be aware of it. As far as the mind, Consciousness, Soul, and Spirit are concerned there is no such thing as time and space either. The immense distances between worlds in our universe present no barrier to the Human Real Self. The "Law of Attractions" leads each person in the proper direction at birth.

These worlds with extraterrestrial life are what many NASA scientists and astronomers have been searching for with no success. Carl Sagan was a notable astronomer who spent a great deal of energy on the quest to find extraterrestrial life. He seems to have known that there is life out there in the heavens but just could not prove it by the scientific mean at his disposal. From the physical standpoint cosmic distances are insurmountable. But if these scientists ever find extraterrestrial life, they will find that it is basically the same as what we have in our earth today, with the exception that they are at various stages of evolutionary development.

And if they ever find them I hope they do not try to go conquer them with the idea to wipe them out. They will not find gruesome forms of intelligent beings as depicted in "Star Wars" and other movies for instance. There are no galactic wars in progress as depicted by many science fiction books and movies.

These worlds did not form accidentally or by chance. They were created to provide a vehicle for Consciousnesses to evolve simultaneously at all levels of development. These worlds are all different from one another yet, they all have the same basic ingredients for our kind of life to exist and flourish. An analogy is that we build homes to live in and they all differ widely from one another, yet all homes have the same basic characteristics, such as kitchen, bad room, bathroom, living room and dining room.

9.5.1 The Odds of Contacting Other Worlds

There is an organization of scientists that use radio signaling to try to contact extra terrestrial intelligences from other worlds. The name of this organization is Search for Extra Terrestrial Intelligence (SETI). What they do is send out powerful radio messages in the direction of various galaxies in the hope that there are intelligences out there – somewhere – that have the know-how to interpreter those messages and then also have the know-how to be able to return some kind of intelligent message back to us. These radio signals have to travel enormous distances at the speed of light; it takes the radio signal in the order of hundreds or even thousands of light-years to travel one way. There is nothing wrong with trying to make contact but the chances of that happening are miniscule.

The odds are unfavorable because the window of opportunity is relatively short. That is, the duration of the travel time of the signal is longer than the duration when the human is technically likely to send and receive these signals. The technical window of opportunity for the human is about 300 to 500 years. The reason for this is that we are at an accelerated rate of development and we will probably be transitioning to the Soul Consciousness Stage in the next 300 to 500 years. Once we get to the Soul Consciousness Stage we will not have a need to look for extraterrestrials because we will already automatically have the capability to mentally communicate with other intelligences anywhere in the universe.

9.5.2 Life

Life is not just a combination of chemical elements such as hydrocarbons, proteins, hormones, H_2O, and DNA. On earth, life is all that plus energy, mind, and Consciousness. In the Spiritual

Domains live is reduced to mind, energy and Consciousness. Life is present in huge quantities in many worlds of many galaxies. Life exists in the billions upon billions and it exists in every level of development from the least significant to the most advanced. All life is sacred and has a Divine purpose to develop and evolve forever forward.

9.6 Cosmic Consciousness Stage

The Cosmic Consciousness Stage is the next higher stage of Consciousness, and many of the characteristics that were mentioned for the Soul Consciousness Stage also apply to this stage. Their evolutionary progress is all done in SD2. They are active in many endeavors that are unimaginable to us. Just because they are very advanced does not mean that they can slack off, relax and do nothing. Even at these superior levels of Consciousness these people are still ignorant of things that lie further ahead on their path of progress. They still have many "veils of ignorance" to overcome, similar in some ways to the way we overcome ignorance in our present life on earth. These folks are extremely intelligent and there lives are characterized as bliss.

Many of these Consciousnesses are always ready to help us in times of crisis. Many will interject and help in Spirit form without incarnating and without us knowing. Many people of the Cosmic Consciousness Stage come to our aid in time of crises such as war or natural disaster. They come to help in times of much trauma, confusion, distress, and suffering. They come to provide comfort, to sooth, to explain and to assure that everything will be OK. People on earth that have gone through the *"near death experience"* report the presence of a *"white light"* that seems to project kindness, comfort, and reassurance. This "white light" of kindness comes from people of the Cosmic Consciousness Stage.

These people have many duties to perform is SD2 and quit a few of them take it upon themselves to help us. A few of these people may incarnate and live among us to help us during a time of crisis. Whenever these people come to live on earth among us they, for the most part, keep a low profile and remain anonymous. But, sometimes some illumined beings may see fit to expose themselves for the purpose of teaching or other worthy endeavor. The upper levels of the Cosmic Consciousness Stage are where many of the Biblical Saints and Prophets come from. This was the case with such illuminating Consciousnesses as Jesus Christ, Mohammad, and Buddha. They came to teach and ended up launched the three most dominant religions that are still widespread throughout the world today.

People of the Cosmic Consciousness Stage have awareness that they are a microcosm of the universe. This means that they think of themselves as being a small representative piece of the whole universe, with this small piece having everything that the universe has in miniature. An analogy of this principle is the human cell, which is a microcosm of the human body. The cell has within it all the information required to clone another human body; this information is in miniature form in the DNA code. People of the Cosmic Consciousness Stage think of the universe as being alive, vibrant, intelligent and manifested by God the Absolute. These advanced Consciousnesses are the Angels, Archangels, Guides, Teachers, Counselors, Saints, and Prophets. They are always willing to help us on the path, and they only require that we be ready for the help. Generally, they know when we are ready when we acknowledge their existence and ask for their help through meditation and prayer. These advanced Consciousnesses will always hear our requests and their help will come in subtle ways through intuition. This help comes so subtly that usually we do not recognize where the help came from. They have been helping us this way for eons, never asking or expecting any acknowledgment or praise. They do it out of love, which is called unconditional love because they expect nothing in return. It is our destiny to eventually become aware of where this help is coming from and we will eventually learn to acknowledge them for their devotion to us.

9.6.1 The lower Emotions are converted to Love

The Soul Consciousness Stage conquered the lower animalistic emotions, which plague us at the Human Consciousness Stage. They took all those emotions to the positive side of the scale. The Cosmic Consciousness Stage now takes those emotions from the positive poles to even higher levels. The higher level above the positive pole is LOVE. Love is actually a complex phenomenon that humans at our level of Consciousness only partially understand. At our level we tend to think that one aspect of love fits all; the English Language has only one word for love. There are languages that have several words for love to depict the several categories or aspects of love.

In actuality there are many types of love and each type is represented by a different vibrating frequency. For instance, we have the love of a mother for a child, the love of a child for a mother, the love of a child for a father, the love of siblings (brothers and sisters), the love of best friends, the love of lovers, the love of husband and wife and so on. All these types of love are similar but they are also different. Each type of love is represented by a "Scale of Love" and each scale has its own range of vibrating frequency. The frequency rate within the scale represents the intensity of a specific love.

As stated, the negative pole and positive pole of an emotion is represented by correspondingly lower and higher frequencies on the scale of the emotion. People of the Cosmic Consciousness Stage carry each of these emotions to even higher levels than the positive pole. They carry each emotion to the level of LOVE, which is a higher frequency than the positive pole of that emotion. To explain this farther, consider the emotion scale of *"cruelty to kindness"*, where *cruelty* is the negative pole (represented by a lower frequency) and *kindness* is the positive pole (represented by a higher frequency) of the same emotion. The people of the Cosmic Consciousness Stage surpass the positive pole of that emotion and go even higher. The higher level of the emotion *kindness* is the *"love of kindness"*, which is represented by a higher frequency of the same scale.

By the same token, all other emotions are now carried to the level of love. For instance, compassion is carried to the "love of compassion", peace is carried to the "love of peace", joy is carried to the "love of joy", and so on. All of these loves are of different types and each carries its own distinct vibrating frequency, which is recognizable by all of the advanced Consciousnesses. This book uses the naming convention *"love of kindness"* for lack of better words. It is appropriate to invent better names at a later date.

9.7 Spiritual Consciousness Stage

The Spiritual Consciousness Stage is the highest Stage of Consciousness in our universe and resides in Spiritual Domain 3. Because it is so far above our present level of understanding we cannot comprehend its glory and significance. We can only give crude approximations of their lofty existence. Information on the Spiritual Consciousness Stages is skimpy and comes down to us from Spiritual Guides who related their sketchy information through advanced mediums and psychics of the past. They can only relate what is understandable to us, but most is incomprehensible; therefore, what they tell us is sketchy and incomplete.

9.7.1 Characteristics of People of the Spiritual Consciousness Stage

The people of the Spiritual Consciousness Stage are approaching absoluteness. They know what absolute truth is, they know what absolute happiness is, and they know what absolute goodness is. They look at our relative world from their vantage point and see it as an illusion, yet they know it is a critical part of everyone's development, and that is as it should be.

These Consciousnesses live in total bliss in Spiritual Domain 3. They live in total peace, harmony, happiness, love, joy, etc. so in affect they live in a Perfect Heaven. *DIVINE LOVE* is their forte or routine way of life. Their lives are so wondrous and wonderful that even if we could comprehend we would not have words to describe or convey. We can only give clichés that are far from true descriptions. Yet even at these heights of existence, they are still active and evolving to even greater states of being. They have high-level jobs and duties to perform, many of which are helping us at our level of development.

There are billions of people that belong to the Spiritual Consciousness Stage and live in millions of planes of SD3. The same hierichical principles and Universal Laws that apply throughout the universe also apply to SD3.

9.7.2 Angels

It is people in the Spiritual Consciousness Stage that are the Angels and Archangels of the highest degree; they are higher than the Angels of the Cosmic Consciousness Stage. Some of these Consciousnesses perform the functions of Spiritual Teachers, Spiritual Guides, Counselors, and Advisors for us in the earth plane. They help us through mental processes such as guidance through mental intuition. They never give us any more than our level of Consciousness can handle or understand. They have a similar function as our personal Soul has in guiding our evolutionary progress. Our Personal Soul has the function of guiding us through our growth process but can only advise us according to our level of understanding at any particular time in our growth. In a similar manner, the advanced Spiritual Guides can provide help to our evolving Consciousness, but only to the extent that it does not exceed the level of our understanding and capability to learn. In essence, Advanced Consciousnesses intervene constantly with our evolution and are always available to help us grow. Our evolutionary process is not automatic or haphazard, as many people may think. They help guide our evolution in the proper direction, without interfering with the Universal Laws that were initially set in place by God. In other words, they have to stay within the bounds of the Law. They have our best interest at heart and we are always safe with them; we are never left alone. Much of their help works through our Intuitive Mind, and that is the reason why we should work to develop the latent potential power of our Intuitive Mind. They are very subtle in the way they help us, and always remain invisible and undetectable to us. They help guide our growth much like our own Soul does.

Evolution of Consciousness

One of the many important jobs that they do is helping to implement the Divine Plan that was discussed back in Chapter 4, Section 4.2.2. It is our job in earth life to do work through physical activity because this activity is the cause of other effects. Since our universe is interrelated and interconnected as a result of the Law of Cause and Effect, our actions produce effects on ourselves as well as on the lives of other people. This is the machinery of life in action or the Divine Plan in action. We each have our place in this great plan of life and we must each do our job to the best of our ability. Our lives are not just for our own personal development, but also for the development of the entire human race. It is one of the jobs of the people of the Spiritual Consciousness Stage to implement this Divine Plan by guiding the manifestations of our actions in combination with the Law of Cause and Effect. We have to learn to work to the best of our ability, and learn to accept and trust the guidance from the Higher Spirit, knowing that the outcome will be good because we are in good hands. The workings of the Higher Spirits are too complex for us to understand so we have to learn to accept it on faith.

The question often arises, "If these advanced Consciousnesses are so willing to help us, why do they allow so much sadism, horror, sadness, misery, and unhappiness in our lives"? The answer is that all those horrible things are relative things that we cause ourselves due to our ignorance. They actually prevent a lot of suffering but they have to remain within the bounds of Universal Laws and we have to suffer as a result of our own doings. We have free will and we bring those horrible things upon ourselves due to our own ignorance. We have to cast off our own "veils of ignorance". The Advanced Consciousnesses will help guide us out of horrible situations, but they cannot learn for us. We have to learn on our own. Their help can be much more effective if we learn to pray and consciously ask for that help. We must also learn to let go and trust their guidance.

Our present society is the most violent society of all the stages of human development. That is, sub-stages 4 and 5 of the Human Consciousness Stage, which are the dominant sub-stages in our society at the present time, are the most violent of all stages of humanity. This is obvious in the television news media, world terrorism, world politics, movies, computer games, kid's toys, and many other activities. We are also putting a heavy burden on our global environment. The Advanced Consciousnesses of the Spiritual Consciousness Stage are constantly monitoring our activities and if we ever exceed that certain limit, which may be approaching a point of no return, then they will step in and alter the course in some appropriate way.

9.7.3 Why Don't They Show Themselves to Us?

A question that is often asked by folks is, "Why won't these advanced Consciousnesses make themselves be known to us in the earth plane?" Actually they are there all the time but remain undetectable. The main reasons for this that if they were to try to communicate and explain to us what they do and how they live, we simply would not understand. The level of their Consciousness is so far above our level of comprehension that it would be impossible for us to understand them. They are multidimensional beings and we cannot comprehend multidimensional stuff. A metaphor that can be used to illustrate this principle is that "They can communicate with us, as well as we can communicate with a fish". In other words, the fish cannot understand us any more than we can understand the advanced Consciousness. The reason the fish cannot understand us is because its level of Consciousness is so far below ours that there is absolutely no compatibility. The same principle applies between the people of Spiritual Consciousness Stage and us. If we had the capability to understand what they do and how they live, we would be inclined to think that they were Gods, but they are not Gods they are just like us; they are Consciousnesses like we are Consciousnesses. The difference is that they are much more advanced than we are because they have already gone through the path of evolution that we are now going through. They have completed their development through all the Stages of Consciousness, while we have a long way to go. Our destiny is to eventually reach the level where they are now.

9.7.4 Region of Absoluteness

People of the Spiritual Consciousness Stage are extremely intelligent. At this level they have discard all "veils of ignorance" and they practically know everything within the bounds of our universe. At the highest levels of their stage, they know all there is to know of the universe and they know how to make the universe work for them. For instance, they know all the physics, chemistry, and biology of the Physical Domain as well as that of the Astral Domain, and all the Spiritual Domains. They know all the intricacies of the Universal Laws as well as the contents of Universal Mind, which means they know all that is knowable in the universe. Throughout their journey they continued to work and learn and progress until they reach this final level in their evolution. At this final level they have reached ABSOLUTNESS. When they reach Absoluteness they merge with their own Personal Spirit and Soul who have been guiding the Consciousness all through its evolutionary journey. Now all three become ONE and are ready for the final step, which is to enter God's Realm and merge with God. At that time they exit the universe to enter the Realm of God and they become unknowable to us, who remain in the universe. When they merge with God they become unknowable because God is Infinite; He is Infinite Mind, and He is Absolute. There is not much more we can say because that is all we can know.

8.7.5 Final Union with God

When the union of the Consciousness, Soul, and Spirit occurs they are ready to merge with God the Absolute and enter His unknowable realm. The Consciousness disappears from the universe at that point. We cannot know what happens to the Consciousness after it transitions to the Realm of God; this is what the most advanced and wisest Spiritual Guides and Spiritual Sages relay to us. As far as we are concerned, it is like the Consciousness enters the infamous "black hole". We can assume though, that entering the Realm of God is a glorious occasion. We can also presume that it is the next step forward into a continuous wondrous journey, but we do not know. We do not know, for instance, if the group remains an independent entity after it enters the Realm of God or if it assimilates and dissolves into the God Being or God Spirit. We don't know if it loses its identity. We just have no way of knowing. Any thing we can imagine from that point on is speculation on our part.

For us who remain in this universe, that is the end of the story of the Evolution of Consciousness. This knowledge can also raise many more questions. Questions such as "Why does God go through the trouble of creating us and than making us go through this extraordinary process of evolution"? "Does God create us for his pleasure"? "Do we add to God's database knowledge"? "What happens to us when we enter God's Realm"? We do not have answers. If God is Absolute, The All, All That Is, and He is Complete, than there is nothing we can say to answer the questions. This book will not endeavor to continue to create and try to answer these questions. It will be left to other thinkers to continue. That should keep many thinkers occupied for a long time to come.

It should be worth mentioning that the Sages do report that just prior to exiting the universe and entering God's Realm, the Trinity does take a break and goes into a period of slumber. This is in keeping with the period of slumber that is taken between any major transition period and completion of any major milestone. For instance, it is similar to the slumber after dying on earth, and also the slumber just before starting a new earth-life.

9.8 Analogy of Imaginary Conveyer Belts

This analogy of a "conveyer belt" is used here to try to add clarity to the process of Evolution of Consciousness. Refer to Diagram 5 that depicts the imaginary conveyer belts. Each of the six conveyers represents one stage of Consciousness. The main thing to notice from Diagram 5 is that Evolution of Consciousness is continuous. That is, new Consciousnesses are continually forming at the beginning or lower-left of the diagram. Consciousnesses progress slowly through the conveyer belt from left to right and after thousands of years they complete the journey at the right of the conveyer, and eventually they transition on to the next higher conveyer belt. This process goes on continuously from beginning of time to the end of time. It has been going on for millions of years in the past and is continuing into the future as far as we can possibly imagine.

Notes for Diagram 5:
(1) The arrows show the path and direction of evolution.
(2) The little arrows represent increments of forward development up the conveyer belt.
(3) Plants and Animals have Group Consciousness that move up from species to species.
(4) The Human Consciousness moves up from level to level using the same basic model of human body.

9.8.1 Plant Consciousness Stage

All Consciousness starts at the lower-left corner of the diagram (figuratively). At this beginning there is a Consciousness generator that produces new Consciousnesses in a continuous manner as if by magic. It is like a "Divine Vaporous Plume" that is emanated by the breath of God. Actually it is not magic; it is like

Diagram 5
Imaginary Conveyer Belt

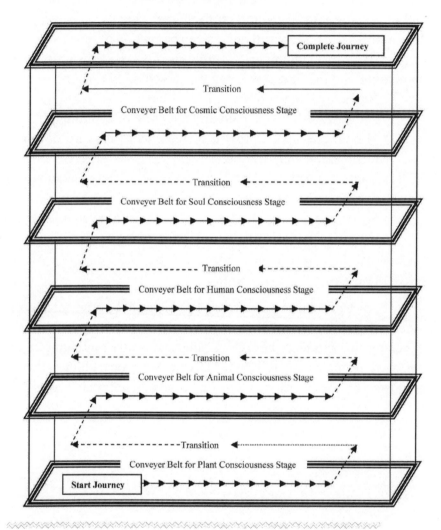

the breath of the Divine continuously spawning these new Consciousnesses. This metaphor is the best we can do to try to give a glimpse of the process.

A graphic that may help visualize this process is to watch the weather report on television where sometimes you can see a spot on the weather map that is generating clouds. It looks like a spontaneous generator that occurs as if by magic, but it is not magic, the weatherman has a complete scientific explanation for it.

At the starting point, the Group Consciousness consists of zero awareness and is completely insignificant, but it is indeed a Consciousness and it is indeed a start. This insignificant Group Consciousness progresses slowly through the Plant Kingdom from one species to the next higher species. Each plant species is a stepping-stone for the Group Consciousness on its way up. The little arrows represent Group Consciousness progressing forward from one plant species to the next higher plant species. The far right of the Conveyer Belt represents the completion of the journey of evolution through the Plant Kingdom. At that point, Plant Group Consciousness is ready to transition to the more advanced Animal Consciousness Stage.

9.8.2 Animal Consciousness Stage

Animals also have Group Consciousnesses. The lowest levels of the Animal Group Consciousness are not much more than the Plant Group Consciousness, yet it represents a quantum leap forward, because the animal has more advanced tools for further development. Animals have a brain with a nervous system and mobility, which plants do not have. This mobility gives the animals a tremendous advantage because it gives them the potential for greater activity, which translates into more experiences and therefore further development. The Animal Group Consciousness also advances from specie to higher species. The Animal Group Consciousness advances from mere sensation in the lower animals, to awareness of the physical bodies in the advanced mammals.

The small arrows represent the Animal Group Consciousness moving forward from one animal species to the next higher animal species. Each species act as a stepping-stone for the growth of the Group Consciousness. When the Animal Group Consciousness reaches the far right of the Conveyer Belt it has reached the highest possible point of its progress, at which time it is ready to transition to the Human Consciousness Stage.

9.8.3 Conveyer Belt for Human Consciousness Stage

When the Animal Group Consciousness advances to the highest possible level within the Animal Kingdom a chunk of it breaks away and transitions to a human. This is a transition that represents a quantum leap. The New Consciousness is born directly into the body of a human baby to start its evolution as a human. As far as the Consciousness is concerned, there is no intermediate body or missing link from the animal to human transition. The transition is direct. It is indeed a profound and extremely significant advancement forward; it is the beginning of the Human Consciousness Stage. It must be realized that these transitions from animal to human are being made incessantly somewhere in our universe, and the progress up the evolutionary Conveyer Belt is also continuously moving forward. There is always someone getting on board at the beginning of the Conveyer Belt and someone finishing at the other end. The transitioning is not a one-time thing as normally thought; it is always and incessantly ongoing.

9.8.4 Conveyer Belt for Soul Consciousness Stage

The transition from the Human Consciousness Stage to the Soul Consciousness Stage is more subtle than the previous transition from animal to human. This transition is gradual; it usually takes one full lifetime or maybe even several lifetimes to complete, and there is no change in body type. That is, the human body form remains the same for the two stages. Each individual Consciousness moves the length of the conveyer as represented by the little arrows. Humans move up from sub-stage to higher sub-stage and transition when they reach the end of the conveyer.

9.8.5 Conveyer Belt for people of the Cosmic Consciousness Stage

The transition from Soul Consciousness Stage to Cosmic Consciousness Stage is also gradual. There is always a transition boundary or a blending between Stages. Each individual Consciousness moves the length of the conveyer one level at a time as represented by the little arrows. When the Consciousness reaches the far right of this conveyer belt it transitions to the Spiritual Consciousness Stage.

9.8.6 Conveyer Belt for people of the Spiritual Consciousness Stage

The people of the Spiritual Consciousness Stage are the highest forms of Human Consciousness in our universe. They start at the beginning of the Conveyer Belt and progress level-by-level up the scale of evolution for thousands of earth years. At the end of this segment of the Conveyer Belt they complete their evolutionary journey where they reach the level of absoluteness and are ready to merge with God. At that time they enter God's Realm and leave the universe forever.

Chapter 10

Techniques for Spiritual Growth

Evolution of Consciousness is the primary purpose for living. This is true whether we know it or not, or whether we like it or not. We do not have any choice in this matter. The only choice we have is whether we want to grow slow or whether we want to grow at a faster rate. Human development has been slow throughout our past history. In recent years that development has accelerated and each individual has a say as to the speed of his or her development. The faster a person develops the sooner that person gets out of this world of unrest and move to a more harmonious life. It behooves each individual to try to accelerate his or her rate of progress.

This book recommends four methods or techniques that will help accelerate the rate of development. These are only recommendations and it is up to the individual whether he or she is interested in using them. These methods can be used singly or in combinations, some people may wish to use all four. The choice is left up to the individual person and depends on the individual's preferences, aptitudes, desires, or beliefs. Actually, all people are already applying one or more of these methods in varying degrees in their daily lives even if they may not be aware of it.

10.1 Four Methods or Techniques for Growth

The four techniques that are recommended to aid in the process of growth are as follows:

1. **Right Living** – Consists of hard work for the sake of work, clean living and devotion to family and community.
2. **Intellectual Advancement** – Consists of intellectual pursuits through education, training, and philosophical endeavors.
3. **Love of God** – Consists of religious pursuits, theological studies, and the love of God through prayer or meditation.
4. **Mind Control** – Consists of conscious control of thoughts, emotions, and habits. Also important is conscious aware of the True Self. This control can be achieved through meditation.

There is no method that is better than another. The best method for an individual to pursue depends on the aptitudes and desires of the individual. There is no right or wrong way. Every individual is different and everyone should apply the level that is comfortable to him or her. It is OK to apply determined effort but not excessive; be balanced and do not become a fanatic in your resolve.

10.1.1 Ignorance

The process of moving up the ladder of unfoldment is to gain experiences that will remove ignorance and propel us forward. Ignorance, as used in this book, should not be construed as a derogatory implication. Ignorance is a lack of knowledge that keeps us from advancing. When we started out in primitive times we knew nothing, now we are fairly well advanced, but we still have much to learn. As we advance in knowledge and awareness, we find that there is always more to learn. It is as if we find that there is always another mountain to climb and it tends to surprise us because we were totally unaware of it. Every step of the way we have to face new situations with total ignorance and every time we have to conquer the situation and move forward to the next situation. This kind of scenario goes on and on as we continually advance forward.

In actuality everything in this universe is knowable and has been known from the beginning. Evolution of Consciousness is the situation in which God created us knowing nothing, but in which he furnished us with the tools to learn everything one step at a time. These tools include our physical body, mind and the *"earth-school"* in which we live; they also include the vast universe with its Universal Mind, and Universal Laws.

We should recognize that from God's standpoint, our Consciousness is the most important essence in the universe; we are more essential than the universe itself. You may ask, "How can that be? The universe is immense while we are small and puny." One answer to this question is that the universe operates on instinctive mind and does not have Consciousness. God created this vast universe, for the singular purpose to provide a means for us to grow in. The universe is not eternal; actually it is relative and therefore an illusion, whereas the Consciousness is real and eternal. We must be aware that we are the most important entities in God's universe and know that He loves us deeply.

Humans are not evil people by nature; we are only ignorant and we carry a lot of trash from past lives. Nature, through its Law of Cause and Effect guides us along the path. A simple example is that of a young child who is warned not to touch the hot grill but out of curiosity and ignorance the child touches the hot grill and gets burned. From this experience the child learns the vital lesson. Evolution is the result of thousands upon thousands of these types of experiences learned over a lifetime or over many lifetimes.

The four methods listed above will help to accelerate the rate of development if used consciously and judiciously. Remember to keep your balance, never overdo it by becoming an extreme zealot. An explanation of each method is offered next.

10.2 Right Living

Action is our natural heritage. All activities contribute to our growth weather it is done rightly or wrongly. When we do things rightly we advance up a notch, when we do things wrongly we

stumble and fall, only to get up again, and try again until it is done rightly and then we advance up a notch anyway. We have been active throughout our past history and that activity has led us to where we are now. This has been so, even though we have gone about it in a haphazard sort of way, and we have been oblivious as to the outcome. Right Living is the technique of development that pursues right action as a means of advancement and it rests on three fundamentals as follows:
1. Hard work
2. Clean living
3. Devotion to family and community

This method of development does not require that the person be religious, or that the person believe in God, or in the Soul, or Consciousness, or in the Spirit. The person can be an agnostic or atheist, but if he or she follows the rules of "Right Living" this person can advance in Consciousness whether the person is aware of it or not.

10.2.1 Hard Work

The idea of "hard work" is to work for the sake of work and not necessarily for the sake of wealth, power, or recognition. Of course, if you work hard you will probably gain wealth, power and recognition without even trying, and that is OK. But, the importance of work is that through work we learn things and we gain experiences. The type of work done does not matter; work is to be enjoyed, but do not be a workaholic. Always try to seek the correct balance of work and rest. The type of work could be physical blue-collar work like driving a bulldozer or it could be intellectual white-collar work as in an office environment. The type of work does not matter as much as the experiences gained as a result of the work done. Work produces experiences, which removes "veils of ignorance". Experiences produce knowledge, wisdom and forward progress. It takes thousands upon thousands of little experiences to create small steps forward towards new level of Consciousness.

Idleness is wasteful. If you are idle and/or lazy you will not make mistakes, but neither will you gain experiences and move forward. Wasting your time does not hurt anything except that it takes you longer to advance. Criminal activity or bad habits such as addiction to drugs or alcohol also contributes to slow growth and is due to ignorance, it is also in opposition to "right living". Moreover these sorts of activities produce adverse experiences, which must be balanced out sooner or later.

Sports and the military produce little in tangible goods, but their training programs are exceptional in producing high quality experiences. The drawback with military training is that much of it is directed at teaching the soldier how to kill his fellow human beings, and that is no good. Most sports including professional sports require intense training that develops the skill for the sport. It also develops personal discipline, physical endurance, mental sharpness, and other traits that are essential to growth. Usually the more intense the training is, the more the benefit.

Such professions as carpentry, mechanics, machinists, educators, farmers, rancher, biologists, engineers, scientists, accountants, business endeavors, and many more are all excellent for producing quality experiences. Hobbies are also beneficial, especially those hobbies or jobs that require meticulous and precise work. Meticulous work such as designing precision machinery or working with equipment that requires a high degree of exactness is also very beneficial.

10.2.2 Clean Living

Clean living refers to keeping the physical body and mind fit in every respect. Some well-known principles to follow for clean living are as follows:

- Proper nutrition,
- Proper physical exercise,
- Proper sleep, rest and relaxation,
- Practice good hygiene and cleanliness,

- Drink clean water and breath clean fresh air,
- Avoid bad habits or addictive influences such as usage of tobacco, alcohol, illicit drugs, and other detrimental addictions,
- Avoid contact with toxic waste products or pollutants,
- Avoid excesses in food consumption especially too much processed carbohydrates such as sugars or white flour products.

All modern medical or health practitioners and wellness programs advocate these guidelines. A prudent and proper balance of these guidelines is essential. It should be noted that fresh fruits and vegetables are better than cooked, canned, or frozen fruits and vegetables. Fresh fruits and vegetables contain more and better balances of vitamins, minerals, enzymes, proteins, and many other nutritional substances, but most importantly, they contain more *vital energy*, which is essential to good health. Vital energy is also abundant in clean air and clean oxygenated water. Stagnant air and stagnant water are poorly supplied with vital energy.

A person that is healthy has an abundance of vital energy. The principles mentioned above will insure high levels of energy. It is normal for young people, especially adolescents and teen-agers, to maintain high energy levels, but older people are naturally depleted of this energy and require more rigorous adherence to the principles in order to maintain as much vital energy as possible.

10.2.3 Devotion to Family and Community

Family ties are central to the development of the human. Humans are social animals for a reason. The reason being that society helps in the development of the person; it helps to propel us forward. Ideally, a person should have strong ties, devotion, responsibility, and love to the immediate family such as father, mother, brothers, sisters, aunts, uncles, cousins, grandparents and the like. The family provides the first line of development, since the early learning first takes place in the family environment. Things such as family values,

Evolution of Consciousness

respect, guidance, discipline, and love are learned from the family. Security, protection, comfort, ethics, and work skills are also first learned at home. All these family learning experiences are invaluable for the growth and evolution of each individual.

As the person matures, more and more learning experiences start to come from friends and community. The principles presented in the book "It Takes a Village" by Senator Hillary Rodham Clinton are true in that the community and society around us help us grow and develop on the path of unfoldment. The schools, churches, and community facilities such as libraries, parks, hospitals, police and fire protection all goes to helping us grow. Every individual's life in our society is interwoven with every other individual. Consequently the actions of one person affect other within the society. Thoughts and actions are what we are all about and they contribute to the influences of the Divine Plan. By the same token, it should be a part of our responsibility to contribute to community functions by participating in their support through voluntarism and/or participation in local government.

When a person reaches maturity, the normal next step is to find a mate, get married and start a new family. This process is also essential to human evolution. The process itself teaches responsibility to both partners in the marriage, and provides priceless experiences needed for growth. The lessons learned growing up are now applied to the growth and benefit of the new family. The responsibility of raising your own family provides immeasurable lessons for your development.

The job should be handled with true dedication and the person should also relish the joy, satisfaction, and true feeling of purpose. Throughout the process of rearing a child there is no substitute for "love of child". The child should be loved totally and continuously; there can never be too much love, and the benefits are mutual. Love produces a balanced life with joy for the child as well the parents throughout the lifetime of the child. The importance of love cannot be overemphasized. The power of love is a Universal Law because it applies in every domain of the universe. It applies in earth life as well as in all phases of Spiritual life.

We have been using these three fundamentals of "Right Living" since the beginning of our evolutionary journey. That is, we have been doing hard work and we have been devoted to family and community since we were in the early "stone ages". From the beginning we had to work hard to gain our food, shelter and security. Our main function then was to hunt and gather food for our family, our tribe and ourselves. We also had to provide security and care for our family and share in the tribal responsibilities. All these early functions were designed for our learning process and they are still good today even though the functions are more sophisticated now.

10.3 Education and Training

The second method recommended for progress is education and training. This includes formal education or self-education gained through reading, working, traveling or whatever means. It is invaluable to the development of Consciousness. The Western World generally has a good educational system that extends from grade school through graduate studies in collages and universities. The more education a person receives the better. The advantage of formal education and training is that structured instruction guides the student in a more direct and efficient path towards mastery. That is to say, it is faster and easier to learn through education and training than it is through trial and error, also known as the school of hard knocks. Education takes advantage of the know-how of others and removes ignorance in a much smoother and efficient manner; thus it accelerates development.

10.3.1 Education

A good education is one that is broad in scope but also has a main course of interest that is focused and intense. All courses of study are good, but there are subjects that may be more beneficial to growth than others. I tend to recommend an education in the physical sciences, since knowing the physical sciences is like learning about Mother Nature, which is a segment of God's work. The physical sciences include the pure sciences such as physics, chemistry, and biology; also of high quality are geology, cosmology, paleontology and many others. All the hi-tech professional studies such as engineering, computer sciences, medicine, microbiology, pharmacy, etc. are also excellent courses of study. The more science and math a person studies, the better it is since knowing nature is a gigantic step forward to understanding the workings of God. It is proper to think of Mother Nature as being a subset of God. This is so whether the student is aware of it or not, and also it does not matter whether the student believes in God or not.

Non-technical courses of study are also essential and produce equivalent results. The arts such as music, theater, and art are extremely beneficial. It has been stated before in this book that you cannot take wealth with you when you die, but you can surely take intellectuality to the Spiritual Domains. Studies in history and philosophy are also excellent to your development. A study in law, business, MBA or CPA are excellent professional pursuits. Of course, the teaching profession is also extremely important since teaching at every level of the educational spectrum has immense benefits to both the student and the teacher. Also beneficial to self-education is reading books in whatever the points of interest are. Travel and exploration of other cultures, customs, and religions can produce great results. It is good provided the traveler maintains an open mind and is open to new views without being narrow-minded about them.

Government support for education at all levels should be praiseworthy. This is so in the United States as well as other countries throughout the world. Education of the young and of the citizenry in general is vital to the society. It encourages economic prosperity of the individual as well as that of society as a whole, but most significant of all is that it advances the Consciousness. An educated society makes for an affluent society that is vibrant and healthy therefore it behooves the government to invest in education. It is also well that the developed countries help the developing countries of the world to achieve a higher level of education wherever possible. Remember that by helping those at the bottom of the evolutionary scale, we also advance the ones at the top. Bigotry and selfishness is no good and should be set aside.

10.3.2 Metaphysics

A course of study, which is alien to most westerners but is beneficial to learn, is the study of metaphysics. The American Heritage Dictionary defines metaphysics as: *"the branch of philosophy that examines the nature of reality, including the relationship between mind and matter, substance and attribute, fact and value"*. Microsoft Encarta Dictionary defines metaphysics as: **Philosophy of being** – *the branch of philosophy concerned with the study of the nature of being, existence, time and space, and causality*. This book expands the definition to include the study of the mind at all levels, and the study of the philosophy of being. This includes the study of the True Self, which is Spirit, Soul, and Consciousness. It also includes the study of the universe and the Universal Laws (causation and relativity). Metaphysics goes beyond the apparent physical world and explores the unseen; the substance beyond the five senses. Metaphysics attempts to answer questions such as, "Who am I?" "Where do I come from?" "What is the purpose of my existence?" and "What is my destiny?" Metaphysics also explores the question of God. It tries to provide the best possible definition of God, knowing full well that it is impossible to define the infinite with a finite vocabulary and a finite human mind. Metaphysics also explores the workings of Universal Law for the purpose of explaining the workings of the universe from an intellectual and scientific viewpoint. People that study metaphysics are comfortable studying objective matters as well as dealing with subjective or abstract matters. Metaphysics is a philosophy that tries exploring, resolving, and explaining the true reality of the universe and of life itself. The tools for doing this are intellectual agility, information from the Intuitive Mind, and information from the masters of old. Spiritual Teachers and Guides that communicate through certain human mediums also provide valuable Information.

This book provides a fairly comprehensive view of the basics of the study of metaphysics. There is much more to learn, and as a matter of fact it is a never-ending study. Metaphysics is a wonderful course of study, but it is not for everyone. Only those who are advanced enough on the path to appreciate its content should undertake the study. It is easy to tell when you are ready to study metaphysics. You will know when you get that unmitigated urge to learn more about it. It is a feel-good sensation of joy, and once you have it you do not want to stop the search for more knowledge. The urge varies in degrees from person to person; some people may have a mild urge while others may have an intense urge. That degree varies according to the level of advancement of the person. In some cases, previously held dogmas or religions beliefs present a barrier to the study of metaphysics. Often church rules discourage parishioners from studying outside of church teachings thereby creating an obstacle for some people, but once the urge gets strong enough; there is nothing that can stop the determined from pressing forward.

10.3.3 Training

There are thousands of training programs, which are equally beneficial to our development. Sports training provide an excellent source of development when it requires intense workout that develops discipline, teamwork, and responsibility among other things. Kids of all grade levels in school should be urged and helped to participate in as many extracurricular activities as possible since they all contribute to gaining useful experiences. Team sports such as little league, soccer, football and baseball provides great training potential. Organizations such as scouting teach skills like canoeing, sailing, swimming, rowing and many more.

10.3.4 Benefit of Hi-Tech

The advanced technology of our modern society has done wonders in the advancement of "Evolution of Consciousness". Technology has advanced in almost every area of our industrial forte. There have been huge advances in the areas of transportation, communication, computerization, medicine, space, construction, the pure sciences, and many other areas. Hi-Tech has revolutionized information dissemination and consequently it has revolutionized the way we think and act. Hi-Tech has cleared many myths of the past and has advanced our unfoldment. The net effect of Hi-Tech is that it causes a feedback loop that hastens growth. The way this works is that, as we get smarter we develop better technology and as a result of the advancing technology we get smarter. It forms a feedback loop that feeds on itself and tends to accelerate our development. It helps us as individuals as well as our society as a whole.

The newest technology that is upon us at the present time is wireless technology. This Wi-Fi technology is promising to start another information revolution on top of the Internet revolution that is now well established. Already people are able to sync up their laptop computers and personal digital assistants (PDA) to the Internet. The tremendous technological improvements in wireless cell phones allow us to communicate using voice, text, and video images almost anywhere in the world at anytime. New innovative cell phones and other digital gadgets for wireless connections are coming on the market every day. All this new innovations will be contributing to our accelerated rate of growth and unfoldment.

The negative drawback of hi-tech profusion has been the pollution and overpopulation of our planet. There is a limit as to how much our planet can take; it is not limitless, as we once thought. It is our responsibility to curb our excesses because if we don't, Mother Nature, through Her Law of Cause and Effect, will automatically clamp down on us. The result will be some amount of needless pain and suffering on our part.

10.4 Love and Devotion to God

The third method for advancing in Consciousness is through the love and devotion of God by way of prayer and religious pursuits. This includes the study of religious theology and faith, as well as the use of prayer. All religious beliefs are acceptable; each person is attracted to the religious belief that is best suited for the level of his or her religious evolutionary level. Whenever the person outgrows or becomes dissatisfied with the chosen faith then that person should feel free to explore another faith. A person should feel free to search for his or her Spiritual calling. The search should continue until the right faith or credence is found. Often that search is not easy, and it may take years, but when the right faith is found the feeling is good and the devotion towards that faith is sincere. The object is to grow in understanding and union with God the Absolute by the power of Love. To know God is to love Him, and the more we know of Him the more we love Him. Of course, we can never know God completely because He is infinite, and unknowable, but we can approach knowing Him to the best of our ability. As we evolve our understanding of Him will also evolve.

Religious study, of course, is not for everyone. We are all of different temperaments and each individual should seek the path that is the most suitable for his or her needs. The religious path or the path of the "Love of God" is worthy, but it should be balanced with other endeavors and the scope of learning should be wide-ranging. Remember that there are many paths to attainment and all steps are necessary when climbing the ladder. Balance is important, whatever you do, do not become a religious zealot, or a religious fanatic, or an extremist. If your minister tells you that the only way to Heaven is through the teachings of his church, than you better beware. Remember that there are millions upon millions of people in the world, and there many different religious faiths and they also deserve a chance to get to Heaven. Do not ever become intolerant or disrespectful of the beliefs of others; we are all on the path; the only difference is that we are all at different levels. Be normal, act natural, and don't be too self-righteous with that pious look that demonstrates a holier-than-thou attitude. A sense of humility, humor, a smile, and a laugh can do wonders for the attitude.

10.4.1 Evolution of Religion

Humans have a natural "religious instinct" that drives us to search for some form of Divine Being. The human search for a deity started back when humans were still living in caves. At first they were ignorant of their worship, but they should not be blamed for their narrowness and blindness because they were doing the best they could, that is the way of evolution. Unknowingly they were all worshiping the same True Absolute God. They were doing this to satisfy the urge of the "religious instinct". The primitive human, driven by the urge and being unable to think clearly on the subject would worship crude symbols. These early peoples had conceptions of God that seem ridicules to us now. For instance, they would worship stone or wooden idles, or they may have offered sacrifices of human blood on their altar. Their enemies were the enemies of their God as well.

The early human usually had many Gods to represent their human emotions, passions, fears, or joys. They had their God of love, war, peace, sun, moon, rain, agriculture, the mountain, the ocean and more. Their God or Gods usually exhibited the same human emotions and attributes as themselves. For instance, if they hate their enemies then their Gods must also hate their enemies. Their Gods reward their friends and punishes their enemies. They imagine themselves as being chosen and favored by their Gods who will go to battle with them and help them triumph over their enemies. They can justify themselves committing atrocities against men, women, and children in the name of their Gods. Their Gods are bloody and savage because they are bloody and savage themselves. And yet God the Absolute is unchanged; these savage people are worshiping and loving their God in the best way they know. Their enemies, at the same time are also worshiping their own conception of God, and also doing the best they can. Both tribes are in essence worshiping their own conception of God to the best of their abilities. They do not realize that they are in fact worshiping the same God under different names, and that God the Absolute loves both tribes equally as well. God does not punish these early people for being ignorant, not any more than He punishes the wolf for killing his prey for food. Their only punishment, if you want to call it that, is that they have to outgrow their ignorance through the long, tedious, and cumbersome process of evolution. The reference made here to God's punishment is only figuratively since God never punishes anyone personally. The Law of Cause and Effect is the one that provides the seeming punishment, not as punishment but as a teaching instrument.

As the human advances in knowledge and understanding their God, or Gods, also advance becoming a little bit better as time passes. As time passes, their Gods become a little bit kinder and gentler, and a little bit more loving than their conception of God of the past. As the humans develop, they begin to see their Gods more like super-persons, who are bigger and stronger and smarter than them. Usually these super-persons are unseen but live in some unreachable place such as on top of a mountain, or up in space, or on a cloud. A common expression used is "Up in Heaven". I am sure this is in opposition to Dante's Hell which is expressed as "Down in Hell". These Gods possess characteristics, which are similar to human characteristics, except that they are much bigger, stronger and wiser. A black person thinks of God as being black, a white person thinks of God as being white, a Mongolian person thinks of God as being Mongolian, and so on. Each one of these Gods is a patriot to the tribe or country to which they belong, and despises the tribe or country of others. By the same token, only their conception of God is correct and only they will go to heaven while all the others will go to hell. These Gods get jealous and angry, and can manifest hate and envy and can change their minds. These God are to be feared and will take revenge if angered. In essence, these Gods are given all the attributes of the human of low development. Primitive peoples form mental concepts of personal Gods that are only marginally more advanced than they are.

The next stage in the development of religion is the development of a single God, but He is still a personal God that has much the same characteristics of the human. This single God is still unseen and lives somewhere that is unreachable, and may wear a white robe with a long white beard. He also lives "Up in Heaven". This God is always vigilant and is always watching over us to see if we are committing violations of the rules so He can punish us. This God is good if we are good, and he is to be feared if we are bad. Of course, the definitions of good and bad are made by the people involved and usually represent the biased opinions of the clergy. Heavens were invented by the clerics to bribe the people to follow the laws of their creed, and hells were invented to frighten them if the bribe fails. Devils were also invented to frighten parishioners into submission and to account for evil. There are religious denominations today that claim that only their version of devotion to God is correct and all others are condemned.

An example of a Deity that favors the people's desires is the God of American Civil War. During the Civil War both sides prayed for victory assuming that God was on their side. The north thought that their God hated slavery and was willing to kill those who were in favor of slavery. The south thought slavery was their right and privilege and they were willing to kill for their beliefs. Yet, both sides were seeing their God through different points of view and seeing Him as themselves. At the present day, both sides see slavery as a thing of the past and now both sides worship the same God, without even thinking about the evolutionary progress they have made. And yet, God the Absolute has not changed at all.

God the Absolute is unchangeable, He was the same one hundred years ago as He is today as He will be one hundred years from now, but the human conception of Him is always changing; this is so because we are relative beings, and therefore we are also evolutionary beings. As we evolve, our conception of God evolves. A person's perception of God is always a little more advanced than the person. The perception of God in the Old Testament is different than the perception of God in the New Testament. The perception of God of the Christians of a hundred years ago is far different from the perception of God of Christians of today. Yet, God is Absolute and remains unchanged; the change is only in the minds of the evolving men and women of the Church. As we evolve we are able to grasp the higher conception of Deity, and thereby that Deity becomes more wondrous.

This brief history of the human journey through the evolution of religion is not intended as criticism or ridicule. It must be realized that no matter how crude, or cruel, or barbarous the forms of worship, they are all steps necessary in the journey. This evolutionary process is still going on and will continue to go on. We are always growing out of old beliefs and moving into better and more advanced beliefs. It must be noted that religious practices have changed greatly in recent years. Religion has grown grander, broader, and kinder each year. The thirst for knowledge of Spiritual matters has extended the search for fulfillment by many in our generation. We hear less of hell and the devil, or of an angry and vengeful God, and we hear more of a loving God. Many churches are now teaching their congregations to love God instead of to fear Him. Words such as "Instill the fear of God in you" are not as common as they used to be. There is no need to beg God for mercy when He represents nothing but goodness, love, kindness, and compassion for us. There is no need to be "God fearing" or to expect "the wrath of God" because God is good; He is not a punishing God. He is not to be feared or dreaded. God is not a policeman that is constantly watching over us to see if we are breaking the law so He can chastise us.

10.4.2 The Advanced Conception of God

Once a person can get away from the personal conception of God, then that person is on the way to higher attainment. A personal God means a God that looks and acts like us, that is, "We are made in the image of God". If we are naughty (sinful) then we will be punished, and if we are good we will go to Heaven. In many religious dogmas of today, their God still has human emotional frailties. That concept of God is primitive and should be discarded.

You may refer to the definition of God that is offered in Chapter 1, it pretty well depicts the scope of the definition for the purpose of this book. Some of the attributes of God are briefly reiterate here for convenience. God is absolute, infinite, eternal, complete, unknowable, non-human, unchangeable, infallible, perfect, and indefinable; God is omnipotent, omnipresent, and omniscient; God is total love, all knowing, all wisdom, and all good. Because God is absolute and complete, He does not need anything else. The universe is of God and it was created perfectly as a complete system with all the necessary Universal Laws and with no extraneous parts. Only one God is present in this universe; there is no such thing as another deity and there is no such thing as a devil that opposes or competes with God. From the absolute standpoint of God, the universe is good, and there is no such thing as bad. From the relative standpoint of the human, there is good and there is bad, and that is as it should be while we are developing. Our job is to minimize the bad and accentuate the good. God has love in great abundance and it is also our job to tap on to that source of love; we need God's love for our own growth and well being. God's love is abundant and it is free for all to share.

On one occasion, Jesus was asked: "What is the greatest commandment?" and Jesus replied, *"Love the Lord your God with all your heart and with all your Soul and with all your mind and with all your strength. The second is this: `Love your neighbor as yourself"*. These statements are found in Mathew Chapter 22, and in Mark Chapter 12. These are true statements, and we should abide by them. We need to love, worship, and adore God, but there is a paradox that we must understand. God being absolute, infinite and complete does not need anything else; God does not need our love and adoration, He already has everything He could ever want. The answer lies in the fact that even though God does not need our love, we need His, and in order to open up ourselves to His love, we have to love Him with all our Soul, and mind, and strength, as Jesus says. His love is abundant and it is free; all we have to do is open ourselves up to receive it.

To illustrate this principle we can go back to the analogy of the sun. The sun radiates sunshine whose rays are warm and comforting to us, and also supplies us with energy, which is essential to our lives. This sunshine is abundant and it is free. It is available to all whether you are good or bad, or young or old, or rich or poor. It does not matter what your religion is or what your beliefs are, and it does not matter what your nationality or your race is or what part of the world you live in. All you have to do to receive this sunshine is to get out of the house and expose yourself to the sun. In a similar fashion, the love of God is abundant and it is free to all. All you have to do is open yourself up to receive it. * The best way for you to receive God's love is to worship, praise, adore, and love God as Jesus says.

10.4.3 Prayer

The best method to achieve the goal of worshiping God is through prayer. We can again quote Jesus on how to pray from Mathew Chapter 6: *"And when you pray, do not be like the hypocrites, for they love to pray standing in the synagogues and on the street corners to be seen by men. I tell you the truth; they have received their reward in full. But when you pray, go into your room, close the door and pray to your Father, who is unseen. Then your Father, who sees what is done in secret, will reward you. And when you pray, do not keep on babbling like pagans, for they think they will be heard because of their many words. Do not be like them, for your Father knows what you need before you ask him."*

This excerpt is an accurate description of how to pray. You should pray in solitude in a quiet place where there are no disturbances. Praying to God should be like talking to your best friend or the person in whom you can confide in and trust, or the person you are most comfortable with. No matter how secretive you are God already knows what you are going to say before you say it; there is nothing you can say that will surprise Him. You may disbelievingly ask "How can that be; it is not possible?" Well, you have to remember that God is omniscient, which means He knows all, and He is omnipresent, which means He is everywhere. Furthermore, God reads your thoughts. Your next question may be "If God already knows what I am going to ask, why should I have to ask it?" The reason you have to ask is because the process of asking establishes your level of comprehension. You can only ask for that which you can understand. If you ask, it means that you are to the level of understanding and therefore you are also to the level of receiving that which you asked for. You cannot receive that which is above your level of understanding. For instance, a person with a PhD can be expected to ask more complex questions than a junior high student. God would not answer the junior high student in the same way that He would the person with a PhD. God will not provide more than the person can handle according to his or her level of development.

10.4.4 Love

Love exists in an infinite variety of categories and potency in our universe and it is poorly understood in our western world. The English Language has only one word for love and we tend to lump all love into that one category. Actually there are many variations of love with infinite degrees power; it is an extremely powerful force of nature. It is a Universal Law that exists in every part of the universe. A "Scale of Love" is used to represent the many degrees of love for each category of love. This scale extends from lowest forms of love to the highest. The lowest forms of love are classified as animalistic lust for sex and the highest forms of love are classified as Divine Love. The "Scale of Love" represents every degree of love between the lowest to the highest. The degree of the potency of love varies with the degree of advancement of the person. The higher a person is on the "Scale of Evolution of Consciousness", the higher the degree of love that can be realized.

At our level of development we can understand various categories of love such as fatherly love, motherly love, love of a sister or brother, love of a child, love of a best friend, and love of a lover. If you think about these types of love carefully, you will be able to see that there are differences in each kind of love. For instance, you do not love your sister the way you love your mother; you do not love your best friend the way you love your sister; you do not love your child the way you love your father, and so on. There are variations in all of these kinds of loves. When you think of fatherly love you think of a provider of such things as food, shelter, clothing, discipline, and guidance. When you think of motherly love you think more in terms of soothing emotional comfort and nurturing security. When you think of sisterly love you think of support, advice, competitive but at the same time caring affection. When you think of love for a child you think of wonderment, joy, pride, protectiveness, loyalty, and deep affection. When you think of love for a lover, a mate, or a spouse you think of deep affection, companionship, commitment, devotion, but which arouses the reproductive instincts of sex. There is also love of nature, love of country, love of work, love of music, love of sports, and many more. These are examples of love in our society, but there are infinitely more categories, types and degrees of love in the universe. There are many extremely high degrees of love in the universe that are magnificent yet impossible for us to imagine. According to the laws of evolution we will eventually reach those peaks and enjoy those wonderful forms of love. For now, we have to content ourselves with what we have and make the best of it.

Now, let us get back to the subject of praying. As mentioned before, when we pray we offer our love to God who in reality does not really need it, but we pray so that we can open up ourselves to his love. Now that we know that there are many forms of love, which love are you seeking? God can provide all types of love; the rule is that He will only provide the kind of love that you seek. For example, if you are seeking fatherly love, you think of the love of a father when you pray and you will receive fatherly love from God. If you are seeking motherly love, you think of the love of a mother when you pray and you will receive motherly love from God. If you are seeking the love of children, you think the love of children when you pray and you will in turn be provided with the love of children. If you are seeking the love of a lover, you think of the love of a lover when you pray and you will in turn receive from God the love of a lover.

10.4.5 An Example of Praying for Love

Suppose that you have been thinking of your mother lately and you have a certain craving for her love. All you have to do to receive that love is to ask God for it. This is an example (technique) of how you can do that. Go to your quiet place, close your eyes, relax and pray directly to God. That is, you strike a conversation where you do all the talking and He does all the listening. You should start out by showing your adoration to God (this is to set yourself up to the right mood or frame of mind). You may say words such as these "Dearest God, Wonderful God, God of Love, Peace, and Joy. I love you with all my heart. Please listen to my prayer." After this you should be ready to start talking to him as if he was your best friend and you may mentally say words to this effect. "Lately I have been remembering my mother and I have been longing and craving for her love." At this point you might elaborate on some of the things that you miss such as her loving smile, or her soft touch, or her wise advice, or whatever it is that you sincerely miss and need. Then you might say words to this effect "Please fill me with your wonderful motherly love" And, when you say that, you visualize a white light that enfolds your whole body. Hold that visualization for as long as is comfortable (10 to 30 seconds). Every person visualizes differently from all others, all that is required is that you do the best you can. Next you can close your prayer by thanking God in your own words. You can also continue a little longer, and ask Him to help you distribute this love that you are now full of; ask Him to help you distribute that love to your family, friends, coworkers, or anyone else that comes in contact with you. This distribution is done automatically; you do not even have to think about it. After you close your prayer you get up and go about your business remembering that that talk was between you and God and no one else.

10.4.6 Law of Attractions

This scenario is an application of the "Law of Attractions". This law is active all over the universe and it is discussions throughout this book. The "Law of Attractions" is like a common thread that appears through everything in our life, no one can escape it. In this scenario we can see that whatever it is that we pray for is what we attract. By the same token, we can say the same about thinking; whatever we think is what we attract. Both prayer and thinking produces the direction for the course of our life. We are what you think because that is what we attract! For that reason thoughts are very important and we better learn to pay close attention to them.

Thought and prayer are similar phenomena. Thought can range from idle thought to determined thought. Idle thoughts (day dreaming) are of low quality and little energy. Determined thought is when the Consciousness applies energy to the thought and as a result it becomes more authoritative. Praying is done in silence and is made of much finer (higher frequency) substance and can be even more powerful than determined thought. When we pray we are able to reach the higher aspects of our nature. The fundamental frequency of our nature automatically increases when we keep high-quality thoughts in mind. Thoughts of nature's beauty, peace, and harmony will increase our inner frequency. Thoughts of compassion, kindness, and all things positive will also tend to increase our inner frequency. Praying when the Consciousness is at these higher frequencies is by definition more powerful.

10.4.7 Praying for help from your Guardian Angel and Spiritual Guides

We know that God is absolute and complete and consequently prayer does not benefit Him in the least. Yet prayer is of the greatest benefit to us, because through prayer we open up ourselves and become tuned in to the infinite strength, courage, and wisdom of God. Through prayer we come into closer contact with Divine wisdom. Of course prayer has to come from the heart, prayer is not lip service; you do not speak or read words without paying attention to what they mean. Your thought behind the prayer is all-important. Divine wisdom is ours and it is free as is air and sunshine; all we have to do is remove the barriers that we have created. Prayer with the proper mental attitude elevates the vibrating frequency of the Consciousness and allows it to reach grater heights.

Prayer can also be used to summon your Spiritual Guides, Counselors and Teachers so that you can have a talk with them. The way to do this is to go to your quiet place, close your eyes, relax and pray in the silence. First you try to get your inner frequency up. You do this by first think of something in nature that is harmonious, beautiful, and happy. You can also recite the Lord's Prayer, or a Psalm, or a poem that is beautiful or illuminating to you. Once you are relaxed and in the right frame of mind, you mentally invite your Spiritual Guides, Counselors, and Teachers to come sit besides you in the silence. Visualize them as entering your room through the door and sitting quietly next to you. You may visualize them as having a luminous oval shape that is unassuming but all wise. You should also know that they come from high level such as SD2 and/or SD3. Then you can start talking to them. You do all the talking and they do all the listening. You must remain aware that they are sitting there next to you as you talk. Next you can make your requests, and be confident that they are listening to everything you say. While you talk, think of them, as being your best friends and you can feel free to divulge everything. You should know that they know everything you are going to say before you say it because they know you better than you know yourself, and you should also know that they would keep your secrets. The talk is between you and them, and no one else. You know that they love you and that they have your best interest in mind. When talking to them you can tell them your feelings on whatever subject you want to talk about, or you can request their help or advice on any matter you need. You can also request their help to help others in need. For example, you can request their help to help some loved one who is sick, or who is in some kind of trouble, or whatever.

There are two requirements when asking for help and they are: (1) The request has to be a positive request, in other words, you can not ask them to perform some evil deed because they won't do it. (2) The request must be within the bounds of the "Divine Plan" which must be within the bounds of the "Law of Cause and Effect" as explained in Chapter 4. As it turns out, most requests (at least 95%) are within the bound of the Divine Plan and most get fulfilled.

As you are finishing your prayer session, remember to ask for guidance and protection. You should also make a mental note of what you talked about so you can recall it later. When you are done you get up and go about your daily business without giving it another thought.

The way that your Guides respond is generally inconspicuous or unnoticeable. They make things happen in such a way that you may tend to think that it was a coincidence, or that it was bound to happen anyway. Another way that they respond is through your Intuitive Mind in which case the solution comes into your Intellectual Mind and you may be inclined to think, "Well, it must have been my own idea." So you give yourself all the credit for being smart. But, if you are really smart you should try to recall the talk that you had previously on your prayer session of the previous several days and realize that those happenings were not coincidence but were a result of your talk with your Guides. They really do not care if you acknowledge them or not, but it behooves you to know where that help comes from. It is also beneficial to acknowledge their help and to thank them even though they do not need it. You have to realize that it is you that needs it; it does you good to thank them.

The way of religious worship is not for everyone, but for those who are naturally attracted to it, it is a worthy pursuit. The religious preference does not matter because the preference depends on the level of development of the person, and as the person develops on the path he or she will likely change their preference to meet the demands of their unfoldment. Daily prayer with strong conviction is required, and in time, unnoticeably perhaps, changes begin to take place in your life. These changes represent your unfoldment on the path.

10.5 Mind Control

The goal of this section is to help elevate of the Consciousness by raising the mental awareness of itself. The principles to help accomplish this are as follows:

1. The Consciousness can be elevated through mental control processes by exposing the negative animalistic emotions, passions, and feelings and than converting those negatives to their positive counterparts. We also have to expose and suppress phobias, traumatic anxieties, and bad habits.
2. The Consciousness must become aware of itself. This means that we have to impress upon the Intellectual Mind that the Consciousness is the Real Self, it is the master of the mind, and that the Intellectual Mind must become intimately and constantly aware of the Consciousness.

Mind control is a method of self-discipline, and there are many techniques that can be used. The preferred technique in this book is the "Silva Mind Control Method" which is based on scientific methods that have been research. The "Silva Mind Control Method" has been used successfully in the United States and in other countries since the 1960s.

10.5.1 The "Silva Mind Control Method"

Mind control is not for everyone. It is recommended only for those people who are inclined to use mental gymnastics as a means to advance. Mr. Jose Silva conducted his scientific research on mind control in the 1950s and 1960s. He wrote several books on the subject, and also has audiotapes and CD's for home study and practice. His books, CD's, audio topes, and workshops are excellent as teaching sources and are highly recommended. This book is interested in meditation only from the standpoint of helping the development of the Consciousness. Some areas where Mr. Silva offers help but this book does not cover are listed below.

Table 1
Benefits of the Silva Mind Control Method

- Learn to Relax
- Control Insomnia
- Alleviate Headaches
- Quit Smoking
- Control Weight Lose
- Improve Memory
- Study more efficiently
- Improve Health
- Learn Clairvoyance
- Learn Extrasensory Perception
- Learn to use Intuition
- Learn from Dreams
- Solve Problems
- Learn to make better Business Decisions

10.5.2 Meditation and Levels of the Brain and Mind

Meditation is a form of relaxing, contemplating, focusing, concentrating, or praying. Meditation is a technique that can be learned, and can be used to accomplish many wonderful things with the mind, which is a powerful instrument that should be used to its fullest. Remember that the brain and the mind are two entirely different things. The brain is physical and the mind is non-physical. There is a common misconception that the mind is hardwired into the brain but that is not so. Yet, the mind and the brain are closely associated and connected to each other through an interface mechanism that translates between the physical brain and non-physical mind.

From the Jose Sylva method we learn that there are four phases or levels of mind that we cycle through on a daily basis. These levels correspond to the brain waves that can be seen and categorized by the electroencephalogram (EEG), which amplifies electrical waves that are generated by the brain. Once amplified, the brain waves can be recorded on a strip chart recorder for analysis. The brain produces a multitude of electrical waves, but there is a fundamental wave that is distinguishable from other waves and represents the operating level of the brain and the mind. The physical brain emits electrical signals; whereas, the non-physical mind emits signals in the mental spectrum but are out of range for our physical senses to detect. These brain waves are measured and interpreted as cycles per second (cps or Hertz). Refer to the table below for information on the four levels of mind and brain waves.

1. The **"Bata Level"** refers to the level where the person is wide-awake. At this level the physical body, physical brain, non-physical HIOM, and non-physical Intellectual Mind are most active, but the Consciousness is in a reduced level of activity or state of rest. At the "Beta Level" the Consciousness exerts little control over the Intellectual Mind and becomes semi-inactive during the hustle and bustle of the outer world's daily activity. At this level the fundamental brain wave (as depicted by the EEG) runs between 14 cycles per second and 21 cycles per second.

Table 2
Brain Waves and Levels of Mind

Phases or Levels of Brain-Mind	Brain waves from EEG in hertz or cps	Description
1. Beta Level	14 cps and higher	**Wide-awake Level**, body, brain and mind are most active.
2. Alpha Level	7 cps to 14 cps	**Meditation Level**; body, brain and HIOM are relaxed, and mind is active.
3. Theta Level	4 cps to 7 cps	**Light sleep Level**; Body and brain are relaxed, and mind is in dream state.
4. Delta Level	4 cps and below	**Deep sleep Level**; body, brain, and mind are deeply relaxed (there are no dreams)

2. The **"Alpha Level"** is the meditative phase in which the fundamental rate of the brain wave is between 7 and 14 cycles per second. At this level the body, the brain, and the HIOM are relaxed, while the Intellectual Mind remains active. In the silence of meditation the Consciousness becomes more active and assumes more control over the Intellectual Mind. The Intellectual Mind becomes an intimate thinking instrument of the Consciousness and dutifully obeys every command. At the "Alpha Level" the Consciousness can more effectively exert its WILL and make things happen. The Consciousness prefers peace and quite.
3. The **"Theta Level"** represents light sleep in which the dominant rate of the brain wave is between 4 and 7 cycles per second. At this level the body, brain, HIOM, and Intellectual Mind are relaxed while the Consciousness is active. At the "Theta Level" the Consciousness conducts most of the

dreaming and may even exit the physical body for short durations.
4. The **"Delta Level"** represents deep sleep where all parts of the human go into a deep relaxation mode. Note that the meaning of deep relaxation does not mean total inactivity for the various components. For instance, the physical body when deeply relaxed remains active enough to carry out the vital functions of the body in order to keep the body alive. The same applies to the brain and the HIOM. As a matter of fact, the HIOM conducts much of the body repairs and rejuvenation during the sleeping hours. It is normal for the person to cycles between the Theta and Delta Levels throughout the night. Sleep scientists have discovered that there is Rapid Eye Movement (REM) associated with dreamtime and by monitoring the REM they can determine when dreams are taking place. Thus they can count the number of cycles that a person makes between the Delta and Theta levels during the course of a night's sleep.

The meditative "Alpha Level" is the focus of this discussion. Meditation can start when you reach the "Alpha Level", making sure that you do not cross over into the sleep cycle. Marvelous things can happen when the brain is at "Alpha Level". A simplified explanation of the process for reaching the meditative "Alpha Level" follows.

10.5.3 Reaching the "Alpha Level"

Sit down in a comfortable chair, in a quiet place, where you will not be disturbed. Close your eyes, and relax your body. For the first 30 seconds or so, you just sit and listen to your breathing. The only purpose of listening to the breathing is that most people can use that as a trick to keep the mind from wondering all over the place. Once you feel that you are settled down, start to make mental suggestions to your body as a whole, suggest to your body to relax and calm down (maybe 15 seconds or so). Next you start focusing on each of your body parts, telling each part individually to relax. For instance, start by focusing on your feet for a few seconds and mentally repeat the word "relax" several times, then move your focus to your ankles and do the same, then to your calves, then to your knees, then to your thighs, then to your hips, and continue to relax every part of your body one part at a time. When you are done with this your body should be completely relaxed. If you go to sleep during this procedure it means that you were tired and needed the sleep and that's OK, but that is not the goal of this exercise. The goal is to deeply relax the body but remain mentally alert. More than likely, if your body is deeply relaxed and your mind is alert you are at the recommended "Alpha Level". It is this meditative mode that you are seeking before you can start your exercise to transform yourself one little bit at a time. Reaching the "Alpha Level" of the brain is called "going to level". Next is a suggested exercise that can be used to convert negatives into positives and should be practiced daily to obtain results

10.5.4 Converting Negative Emotions to Positive Emotions

Converting negative emotions and thoughts to positive ones through force of will and self-discipline is an excellent method to advance and even accelerate the process of growth. The technique to use for this purpose is meditation. The net affect of converting emotions and thoughts from negative to positive through meditation is equivalent to circumventing the "Law of Cause and Effect". Definitely an explanation is required here!

The Law of Cause and Effect is a firm Universal Law that is infallible and immutable, but it does allow a seeming loophole. Mother Nature does not care how you learn your lessons as long as you learn them well. If the lessons are learned, nature will forgo the balance of the effect of a cause. Another way of saying this, and still maintain a close approximation of the meaning, is that if you *repent* you do not have to pay for past *sins*. Converting emotions and thoughts from negative to positive is like making a personal transformation. This transformation will change the person from one of low standing to one of advanced standing. What actually happens during the transformation is that the signature frequency of the person moves up the scale on the Scale of Evolution of Consciousness. If the signature frequency moves up the scale sufficiently, it can bypass the frequency of the original infraction, and thus bypass the infraction without having to suffer the consequence of the infraction.

We have already covered the method of advancing through hard work, which produces experiences as well as errors. Many of the errors that we commit are due to the hidden animalistic emotion and feelings that are still engrained in our subconscious (the HIOM). Instead of going through the cumbersome and long-term process of trial and error, we can more easily use meditation and force of will to advance rapidly on the path of attainment.

10.5.5 Emotional Awareness

Animalistic emotions were necessary for the evolving animal to survive in the wild, and became a part of the animal's AIOM. Through evolution the animal emotions transferred to us and remain part of our HIOM. In our primitive sub-stages of development this negative emotions served us well because at that time we lived much like the animal. Now, at our present level of Human Consciousness it is our job to diminish the affects of those primitive emotions because they are harmful to us now. It is now time to remove those ancient emotions, but before we can begin removing them we must first become aware of then.

The old animalistic emotions and passions are what the Bible commonly calls *temptations*. These temptations are powerful because they are ingrained in the old subconscious and unbeknownst to us they seek expression. When the temptations win over the better judgment of the person they become *sins* in the language of religion. The idea is to remove those temptations by force of will, and this elimination of temptations is what religion calls *repentance*. Orthodox religion normally associates temptations as driven by the Devil; this book is vehemently against that notion. Animalistic emotions are normal; they are there for a reason and it is now time to consciously remove them.

All emotions come as "emotional pair" and they consist of the negative emotion (negative pole) and the opposite positive emotion (positive pole). These are two different poles of the same "emotional pair". Below is a list of a few of these pairs.

The emotions listed in the table are only a few of the many emotions that manifest in our being. The negative poles represent the lower part of the emotions and the positive poles represent the higher part of the emotions. For example, "cruelty" is the lower negative pole and "kindness" is the higher positive pole of the same "emotional pair". The "emotional pair" is also represented by an equivalent "emotional scale". A range of vibrating frequencies that extends in degrees from the negative pole to positive pole constructs the "emotional scale". The person's emotional position on the emotional scale is represented by a "frequency marker", which indicates the degree of advancement on the scale.

Adjectives can be used to try to describe the various degrees of the emotions on the "emotional scale". For instance various degrees of cruelty may be expressed by words such as malice, spite, brutality, meanness, and vindictiveness. By the same token degrees of kindness can be expressed by such words as compassion, sympathy, gentleness, helpfulness, and kindheartedness. These are all degrees of the same thing on the same scale; Mother Mature represented these degrees on a "frequency scale". An important criterion of "emotions scales" is that different scales do not mix with each other. In other words, you cannot mix the emotion of hate-love with the emotion of impatience-patience, or with harsh-gentle, and so on.

10.5.6 Becoming Aware of the Emotional Average

Most people swing back and forth through the scale of their emotions. Often a person may have a negative peak and later swing back to the positive peak of the same emotion. Every person has an average on the scale for each emotion, and should become aware or his or her "emotional average". It is perhaps a good idea for an individual to draw a chart of each emotion to determine the average and thereby establishing emotional awareness. A simple chart with a

Table 3
Emotional Pairs and their Polarities

Negative Pole	Positive Pole
Hate	Love
Fear	Courage – Bravery
Anger	Calmness
Timidity	Confidence – Self-assurance
Greed	Restraint – Moderation
Heartless	Merciful
Cruel	Kind
Harsh	Gentleness
Selfish	Unselfish – Generous
Discord - Acrimony	Harmony
Pandemonium	Peace
Confrontation	Consensus
Self-indulgence	Restraint – self control
Impatience	Patience – fortitude
Sorrow	Joy
Blame	Pardon
Falsehood	Honesty – Truthfulness
Grief	Happiness
Bigotry – Prejudice	Fairness – Tolerance
Condescending	Agreeable
Lustful	Modest – Moral

scale of 1 to 10 for each emotion of concern should be sufficient to supply the needed information of emotional awareness. For example draw a line to represent your chart for the "emotion pair", "anger to calm", and write the numbers 1 to 10 on the chart. Then place a mark on the chart that represents the average anger or calmness for the day, or for the week, or for the month.

From this you can see where you stand. If too many of these emotions lean towards the negative side of the chart you can say, "I am a bad person and I have to change my ways." If most of these emotions lean towards the positive side of the chart you can say "I am basically a good person, I am well on the way to unfoldment, but I can always use some improvement."

The next step is to try to make a transformation and change your ways. You can either continue to use Nature's way of trial and error or you can go an easier way of mind control through a form of meditation. Mother Nature does not care how you learn your lessons as long as you learn them well. You might ask, "How can Nature keep track of where on the scale of every emotion, each human in the world stands?" you might say "That is impossible." Of course nothing is impossible for Mother Nature. Her design is superb, accurate, fair, and unfailing. The way it works is that every individual, subconsciously, keeps track of his or her own emotions by automatically placing a "frequency marker" on the "emotional scale". Mother Nature only has to monitor that "frequency marker" to know where you stand.

Your job now is to become aware of your negative emotions and then move the "frequency marker" up the "emotional scale". That "frequency marker" is like a beacon, which is accurate to the nth degree with no chance of fraud are abuse. There is no way to cheat, and the only way to move it forward is "the old fashion way, you have to earn it".

10.5.7 Back to Meditation

Now we can get back to the subject of meditation as a means of moving that "frequency marker" forward on the "emotional scale". Before you sit down to meditate, and while you are still wide-awake, you should review your list of charts. From that review select one, two, or even three of those emotions that you feel you need to work on. After you make your selection you should be ready to sit down to meditate. To begin your meditation, go to your quiet place and inter your "Alpha Level".

Once you feel confident that you are at your "Alpha Level", you are ready to begin, as follows. Suppose that you selected the anger-calmness emotion to work on. Then you mentally make up your own affirmation that goes something like this, "My emotional awareness chart indicates that I get angry easily and I tend to throw temper tantrums. I know that anger is detrimental to me and to those around me. There is no reason for that kind of behavior. I do not want to be angry. I want to be calm. I will be calm. I will be calm, tranquil, soothing and peaceful." You repeat this or similar statement several times, making sure that you are committed and sincere about what you say.

Before you end your meditation session you should always add an affirmation or prayer for protection. You can make your own and it should go something like this, "God is present in my home and in my life and He will protect me from all evil." At this time you may also pray to your Guardian Angle or Spiritual Counselor and ask for guidance and protection as well. Use the same basic procedure as above for each emotion. A meditative session should last around 15 to 25 minutes.

When this is over, gradually waking up, try to forget what you did, and go back to your normal business. This exercise should be performed every day at least once a day, but preferably twice a day. There is no harm in meditating three or more times a day. With each meditation session, you work on an emotion that you feel is giving the most trouble. Over a period of time you cover quite a few of these emotions and the end result is that you improve your emotional structure and thus become a better all around person.

In essence, what is happening is that you are raising your level of Consciousness, by raising the rate of frequency for each emotion. When you say, "I want to be courageous and brave." or you say "I want to be loving." or you say "I want to be kind and compassionate." or you say "I want to be peaceful and harmonious." then you raise the frequency of those emotions by a notch. This frequency increase is automatic and it is cumulative; it is done in your subconscious and you are unaware of it. That is the way Nature keeps track of where, on the scale, each person is. It must be noted that the negative emotions will never be completely erased from your subconscious. Those negative emotions have existed in the instinctive mental makeup for millions of years and will never go away completely. They will however be suppressed and sent to the background to the point where they become insignificant in your everyday life, and the positive emotions will instead rule your life.

Because the negative emotions remain imbedded in the instinctive subconscious, they can occasionally appear unexpectedly and come back to haunt you. You have to remain alert of these possibilities so you can take corrective action if the problem occurs. The consequence of suppressing the negative emotions and shifting your overall emotional vibrations to higher frequencies is enormous. If you make an effort to remain alert, you will over a period of time, become aware of the personality change. The benefits are much greater than you can see at first glance.

10.5.8 Phobias

Phobias belong to the same basic category of problems as animalistic emotions. They need to be removed from the foreground of the subconscious and sent to the background where they do no harm. The difference between animalistic emotions and phobias is that animalistic emotions are ancient while phobias are fears that have been acquired in the present life on earth or in recent past lives. A few of the phobias that are common among many people are:

- Claustrophobia – fear of enclosed places
- Ophiciophobia – fear of snakes
- Ochlophobia – fear of crowds
- Pyrophobia – fear of fire
- Brantophobia – fear of thunder

These are only a few of the many phobias that are common in varying degrees. We only have to worry about those that are problematic or unmanageable. "Posttraumatic Stress Disorders" and many "anxiety disorders" are similar types of mental problems as phobias. Phobias normally come to us from past lives whereas Post-traumatic Stress Disorders are products of the present life. Psychiatrists have developed excellent treatments for many kinds of phobias and other mental disorders. These treatments should be used wherever possible but often they are expensive and many people cannot afford them.

It is important to differentiate between physical brain problems and non-physical mental problems; they seem to have similar symptoms but they are different and should be treated differently. Note that animalistic emotions and phobias are mental problems that have nothing to do with the physical brain. Physical brain problems may pertain to chemical imbalances or some kind of physical damage or degradation of the brain. Degradations of the brain include Alzheimer's and Parkinson's.

The meditation technique for removing phobias and other mental disorders is similar to the one mentioned above for removing animalistic emotions. Suppose that you have a problem with claustrophobia, which is fear of enclosed places. To combat this problem with meditation where you go to your quit place, enter your "Alpha Level" and then you mentally make up your own affirmation that goes something like this, "I have a problem with claustrophobia; it is not normal to have an inordinate fear of enclosed places. Extreme claustrophobia is not normal and I do not want to be hampered by it. I do not want to be afraid of enclosed places and I will not be afraid of enclosed places." You are to repeat this or similar affirmation several times during your meditative session. This or similar techniques can be used for fear of snakes, fear of thunder, fear of rats, fear of fire, fear of heights, or whatever the phobia may be. A similar technique can also be used for "Post Traumatic Stress Disorder".

10.5.9 Bad Habits

Bad habits are also of the same category as animalistic emotions and phobias because they are all ingrained into the subconscious in a similar way. The difference is that bad habits have been acquired in our present life. We acquire bad habits in many different ways; we may get them from family habits, from society or cultural habits and from peer pressure. We create many bad habits through inputs of the Trainable Mind. Bad habits include such things as cigarette smoking, use of excessive alcohol or elicit drugs. One very bad habit that is rampant in our society today is excessive food intake; in my opinion it is as bad as tobacco, alcohol, and drugs. At the present time obesity is a serious problem in our society and the craving to eat is as bad as the craving to smoke or drink. There are thousands of things that can be classified as bad habits, and they all have to be suppressed in a similar fashion as negative emotions and phobias. They all have to be suppressed because they tend to interfere with our forward progress. The Trainable Mind is a wonderful tool but it does not know the difference between good and bad; it learns whatever the Intellectual Mind directs it to learn. Nonetheless, bad habits have to be removed or suppressed through self-discipline and through the technique of meditation.

10.5.10 Benefits

Some of the benefits that can be expected as a result of suppressing animalistic emotions, phobias, and bad habits are as follows:

- On the surface you become a better person. This becomes obvious when all your attributes shifted to high virtues rather than low standards.
- Of greater impact is that you actually climbed several rungs up the ladder of "Evolution of Consciousness".
- You actually circumvent the Law of Cause and Effect. This means that perhaps the effects of some of the bad (negative) infractions that you produced in the past will be canceled out. This is possible because the negative frequency of the infraction can be counterbalanced by the positive frequency of your advancement, and the influence of the infraction will be negated.

- Another enormous benefit to you is the new influence of the "Law of Attractions". When your emotional composition was negative you produced negative thoughts and attracted similar negative thoughts towards you since; "Like Attracts Like". Once you transform your negative emotions to positive emotions, you begin to produce positive thoughts, which in turn attract more positive thoughts towards you. Your over-all attitude now continues to improve at an accelerated rate because your positive thoughts are continually reinforce with more positive thoughts. This is a positive feedback loops that is cumulative and very advantageous.

10.6 Awareness of Consciousness

The previous section dealt with conversion of emotions, phobias, anxieties, and bad habits. This section emphasizes the importance of recognizing that the Consciousness is the Real-Self and is the masters of the mind and body. We must become conscious of the fact that we are Consciousness and that the mind and body are tools to help us advance.

When we see and hear the voice of another person we are actually seeing the animation of the other person's Consciousness. The body of the person would be lifeless without the Consciousness animating it. The Consciousness is hidden in the recesses of the subconscious, and must be brought to the forefront of conscious awareness. The Consciousness is the part that portrays the level of evolution of the person. The Consciousness is a Real Thing; it is us at every step of our evolutionary journey.

Most people do not realize that they are really the Consciousness. Most people think of themselves as being a personality and physical body with the capability to think. A French philosopher and mathematician, Rene Descartes (1596-1650) once wrote, "I think therefore I am." This idea is an immense advancement over the previous perception of the self, but it does not go quite far enough. A more accurate statement would be "I am a Consciousness therefore

I AM." As we advanced we will eventually become conscious of being a Consciousness that has a body and mind, and as we advance further we will become aware that we are the masters of our mind and body. This is when we begin our transition to our next stage of development the *Soul Consciousness Stage*.

The Consciousness is a kin to the mind, but it is superior and master over the mind. The mind is a subordinate to the Consciousness, even though most people do not realize it. It is up to the Consciousness to assert its authority and responsibility, and become dominant over the mind. The Consciousness must bring itself to the forefront of conscious awareness by training the mind to respond. The Consciousness must become aware of the fact that it is the **"center of thought, influence, and power"**, and until it does that, it will not be able to be the true master. Remember that the Consciousness has advanced from Plant Consciousness with zero awareness to Animal Consciousness with awareness of the physical, to Human Consciousness with awareness of the physical and mental. Now it is time to move up to aware of the Spiritual.

The Consciousness is eternal and cannot be harmed or killed by fire, or knives, or bullets. As an extreme illustration of its indestructibility, suppose that someone wanted to obliterate and completely destroy another person because of pure meanness. Let us suppose that the victim is tied to a nuclear bomb and the bomb is detonated. Of course the body would be blown to smithereens, but the Consciousness, who is the real person, will not be affected. The reason this is so, is that the atomic bomb produces physical energy that affects the physical body only. The Consciousness is non-physical; it is made of an entirely different kind of substance; it is made of vibrating wavelets of extremely high frequencies that are incompatible with the physical vibrations. Therefore, the Consciousness is impervious to the physical blast of the manmade bomb. The Consciousness is not of the physical world, it is of the Spiritual world and nothing of the physical can possibly harm it.

You must become aware that the **WILL** comes from the Consciousness. The Consciousness can impose its will on the mind and the mind must obey. Once you learn that you are master of your mind and not its slave, you will be able to be master of your moods and emotions as you see fit. You are the driver and not the driven. The mind has kept you oppressed in the past due to ignorance, but now it is time for you to assert your will over the mind and become boss and be free.

10.6.1 Tools of the Consciousness

The tools of the Consciousness are the body, energy and mind. Tools are temporary and continually being replaced as follows:

Body: The physical body is made of atoms, molecules, proteins, hormones, and living cells, which can be classified as matter in general. Matter is continually flowing in and out of the body; it is constantly being replaced. For instance, all the cells of the skin are completely replaced every thirty days or so, and the cells of most other organs of the body are also replaced periodically in a similar fashion. There is a continual inflow and outflow of matter, in and out of the body. The food we eat, the water we drink, and the air we breathe represent the inflow of matter. The outflow of matter is represented by the elimination of material through urination, perspiration, and breathing. Your body is made of the same matter as the physical universe (Physical Domain). It is a part of the physical universe and is completely immersed in it. It is in constant and intimate contact with the universal ocean of matter. Yet, you are separated from the universal ocean of matter only by the skin that encloses your body.

Energy: Energy works in a similar way as matter. There is a continual inflow and outflow of energy, into and out of the body. You can feel the outflow of energy when you work too much and as a result you get tired. The inflow is when you eat good nutritious food; you drink good clean water; you breathe good fresh air, and you get plenty of rest. You live immersed in an ocean of universal energy. Yet, you have your own personal energy, which is separated from the universal ocean of energy by a thin energy membrane.

Mind: Mind works in a similar way as matter and energy. There is a continual inflow and outflow of mind from the ocean of Universal Mind substance. Your Personal Mind is immersed in this ocean of Universal Mind substance, but is separated from that ocean by a thin mental membrane.

It is fairly simple to understand of the principle of inflow and outflow for matter and energy, but the inflow and outflow of mind is probably more difficult to understand. What is important is to recognize that the body, energy, and mind are tools of the Real Self. The tools are always changing and are not permanent; in contrast the Consciousness is permanent.

10.6.2 Computer Analogy

It is difficult for most people to think of themselves as a Consciousness because it resides in a subconscious realm, but it is now time for many of us to start taking steps to become conscious of the Consciousness. An illustration that might help clarify the mental picture of the Consciousness is to use the computer as an analogy. The computer is a good metaphor in explaining the human being, because there are many similarities between a human and a computer. Refer to the table below, which shows human component parts and the seeming equivalent computer parts.

Table 4
Analogy: Human Components vs. Computer Components

Human Component Parts	Computer Component Parts
God - Creator of the universe	The human - creators of the computer.
Consciousness	Human Operator Sitting at the Keyboard
Intuitive Mind	Internet
Intellectual Mind and Personality	Application Software
Human Instinctive Operating Mind	Operating System
Energy	Power Supply
Brain	Central Processing Unit (CPU)
Physical Body and Astral Body	Computer Hardware

The human creators of the computer, which are the scientists and engineers, can be figuratively equated to God, and the Consciousness can be figuratively equated to the human operator sitting at the keyboard. This is so because the Consciousness is master of the mind and body, and similarly the human operator is master of the hardware and software of the computer. No matter how complete and sophisticated the computer is, it cannot direct itself, the human operator is indispensable to give it direction.

The awakening to the realization that we are indeed a Consciousness that is eternal is the first step in awakening to the realization that we have the power of the universe within us. It has been a long journey and there is still a long way to go but we are now getting close to entering the conscious stage of Spiritual Evolution. From now on we will be able to see the path ahead much more clearly.

10.7 Mastery of the Mind

During the early sub-stages of human development we lived much like the animal, we lived by instinct alone. As we progressed up the scale of evolution the Intellectual Mind gradually became more dominant. At our present level of development most of us live by the wit of our Intellectual Mind, and it is now time to start working towards the next step. It is now time in our evolutionary process to assert the mastery of the Consciousness over the Intellectual Mind and the HIOM. Progress toward the goal can be speeded up by a methodical and systematic approach that can shorten the period of time to transition to Soul Consciousness Stage. Many in our society are now at the threshold of this significant occurrence, and should take advantage of it in order to accelerate their growth. It is our destiny and responsibility to continue to evolve.

The recommended technique to accelerate the progress toward transition is the use of the *Silva Mind Control Method* of meditation along with certain affirmations. The goal is to train the mind, or to program the mind, to become conscious of the Consciousness. Two goals are as follows:
1. Program the Intellectual Mind to think of the Consciousness as the master. The Consciousness is the center of thought, influence, and power.
2. Program the mind to be constantly aware of the Consciousness as being the Real Self and residing inside the body.

10.7.1 Programming the Intellectual Mind

Meditation is used as the technique for programming the Intellectual Mind to think of the Consciousness as master. The general process is to sit in a comfortable chair, close your eyes in a quiet place where you will not be disturbed, and go to the "Alpha Level". Both the HIOM and the Intellectual Mind must be tamed. The HIOM is a wonderful mind but it contains a huge amount of rubbish as well. The Intellectual Mind will tend to ramble uncontrollably like a frisky monkey that is constantly on the move, and it must be slowed down long enough to be trained. The Intellectual Mind is not the Real Self, but it is the mechanism that the Consciousness uses to think. An efficient method to discipline the Intellectual Mind is through the use of affirmations in conjunction with meditation.

Affirmations are also known as mantras or autosuggestions. This book provides several examples of affirmations that may be used as a guide.

10.7.2 Application of Affirmations

Several affirmations are provided here to serve as guides for those who wish to use this method. The object here is to train the Intellectual Mind to settle down through meditation and then train it to know its place in the hierarchy of the Real Self. The idea is to program the Intellectual Mind to be conscious of the Consciousness and to think of it as the Real Self and master. Most of these affirmations are based on affirmations provided in the book "Raja Yoga" by Yogi Ramacharaka. Raja Yoga is a philosophy of the east that deals with mental development.

When you meditate, think of yourself as being a Consciousness that is the center of, power, influence and thought. You are these qualities in the proportion that you are able to realize it, and your abilities to manifest these qualities will also be in the same proportions as you realize it.

When using these affirmations, it is recommended that you concentrate on one affirmation at a time. Before starting the meditative session you should read the affirmation several times and contemplate on it; you may also change the wording to suit your needs if you wish. The affirmation that you choose to use for the session should be almost memorized so you don't have to grope for words during your meditation. Once you feel that you are ready to start, you sit in a comfortable place that is quiet, and you close your eyes. Then you slowly and calmly relax and gradually go to the "Alpha Level". Once you reach the "Alpha Level" you begin to leisurely recite your affirmation with conviction and resolve. This affirmation should be mentally repeated several times during your period at the "Alpha Level" and it should be done slowly, deliberately, thoughtfully, and sincerely. You cannot fake your thoughts; remember that you can never deceive Mother Nature.

Affirmation Number 1

I am a Consciousness and I am the center of thought, influence and power. I am master of my mind and body, not the slave. I am the Real Self; I am independent of the mind and body; I am immortal, invincible, and invulnerable.

Affirmation Number 2

I am a Consciousness. I am a Center. Around me revolves my world.

I am a Center of Thought, Influence, and Power.

I am Independent of my Mind and Body.

I am Invincible and cannot be injured.

I am Immortal and cannot be destroyed.

Affirmation Number 3

"I am a Consciousness – my mind is my instrument of expression.

"I exist independent of my mind, and am not dependent upon it for existence or being.

"I can set aside my sensations, emotions, passions, intellectual faculties, and all the rest of my mental collection of tools as "not I" things.

"My Consciousness is the Real Self, it is the "I AM".

"I am master of my mind; my mind is my tool and is subordinate to me."

Affirmation Number 4

"There is but one ultimate form of matter; one ultimate form of energy; one ultimate form of mind. Matter is created from energy, and energy is created from mind, and all are emanations of God the Absolute, threefold in appearance but one in substance. Life manifests in many forms but in actuality there is but One Life that permeates the universe, it is that of the Absolute. My body is one with universal matter; my vital energy is one with universal energy; my mind is one with Universal Mind; my life is one with Universal

Life. God the Absolute has manifested itself as Spirit, which embrace all the apparently separate "I AM's" and I identify myself with that Spirit. I feel my union with ultimate Spirit through my own Real Self. I realize that I am an expression and manifestation of God the Absolute, and that its very essence is within me. I am filled with Divine Love. I am filled with Divine Power. I am filled with Divine Wisdom. I am conscious of my identity with God the Absolute; the One True and Ultimate Reality."

Affirmation Number 5

"I am Consciousness and I have a WILL, which is my irrefutable property and right. I am determined to cultivate and develop it by practice and exercise. My mind is obedient to my WILL. I assert my WILL over my mind. I am master over my mind and body and I assert that mastery. My WILL is full of force, energy, and power. I am strong. I am forceful. I am vital. I am the center of energy, strength, and power, and I claim my birthright."

Affirmation Number 6

"I am a Consciousness with senses to experience the world. I will use these senses to the obtain information and knowledge necessary for my mental development. I will use my existing senses knowing that training and perseverance will help unfold my latent higher senses. I will be wide-awake, alert and open to the inflow of knowledge and information. The universe is my home – I will explore it."

Affirmation Number 7

"I am a Consciousness that is far greater and grander than I have as yet imagined. I am developing gradually but surely into higher levels of Consciousness. I am continuously moving forward and upward. My goal is the realization of my True Self, and I welcome each stage of development that leads me toward that goal. I am a manifestation of God the Absolute: I AM a CONSCIOUSNESS!"

Affirmation Number 8

"I recognize that I am a Consciousness that is continuously evolving. I realize that there are sub-stages of my Consciousness in my past that I have outgrown and left behind. I also realize that my future development contains wondrous things. I assert my mastery over the lower phases of my mind and I demand of the higher phases of mind all that it has to offer."

Affirmation Number 9

I am a Consciousness and within me there is a great area of subconscious mind that is under my command and subject to my mastery. This mind is friendly to me, it is glad to do my bidding, and obey my orders. It will work for me when I ask it, and it is constant, untiring, and faithful. Knowing this I am no longer afraid, ignorant or uninformed. I am master over my body, mind, and subconscious and I am asserting my authority. I am the center of power, strength, and knowledge. I am also Spirit and Soul, which is a fragment of the "Divine Spark".

Affirmation Number 10

I am master of my mental habits and control my character. I apply my strength, desire, and WILL to affirm my power over my mental habits.

10.7.3 Illustration for Developing an Awareness of Self

It is essential to be aware of the Real Self within you. This awareness should be continuous throughout your entire working day. You should be mentally aware throughout the day that the Soul and Consciousness are quietly residing inside your body in the vicinity of your heart. They appear to be quiet and not interfering with your activities but in reality they are guiding your actions. An illustration that can be used to explain how you should be mentally aware of your Consciousness is as follows:

Suppose that you are driving on a long cross-country trip and you have one passenger with you in the car. However, this passenger is deaf and mute consequently the trip is driven in silence. Even though you do not speak to your passenger you are always completely aware that he is there in the passenger sit. You may be listening to the car radio or having your own personal thoughts while driving, yet you never forget that your passenger is there. If you stop for gasoline you acknowledge that he is there, and you see to it that he goes to the restroom if he wishes, or that he gets food and drink if needed. In other words, even though you do not speak to your passenger you are always aware that he is with you. This example illustrates the idea that you should be continually aware that your Consciousness is always within you in a similar way that you are aware of your silent passenger in the car.

You should always be aware that you, the Real You, the silent Consciousness You, is always within your physical body. Your body is the "Temple of your Consciousness". You should realize that the silence of the Consciousness is only in appearance; in actuality, the Consciousness is always busy. The reason it appears silent is that the Consciousness is of the Spiritual World where there is no such thing as physical sound. On the other hand, you live in the physical world where there is sound and noise. Physical noise is not compatible with the Spiritual World consequently in order to make contact with your Soul-Consciousness you must do it in silent meditation. Silent meditation removes the clutter of the outer world.

Remember that your Soul is basically your primary guide throughout your developmental process. If you become aware and acknowledge that your Soul is within you at all times, then your Soul will be more apt to give you that guidance and help through out the day as you need it. The Soul is always ready, willing, and able to lend a helping hand in all your daily activities; however, it becomes more effective when you consciously acknowledge its existence and life overall will become easier. This is so because you will find yourself making fewer mistakes in judgment since you are getting more help from "Your Higher Self". You have to learn to listen and trust because you are always in good hands.

10.7.4 Acceleration towards Soul Consciousness Stage

The purpose of practicing meditation with affirmations is to try to accelerate the progress towards transitioning to Soul Consciousness Stage. The eventual consequence of the transition will be profound. Some of the changes that can be expected are as follows:

- Develop new and more advanced latent senses that are in addition to the regular five senses that you now have.
- Become conscious of not only the Physical Domain, but also the Astral Domain and Spiritual Domain 1.
- Become conscious of life on earth as well of life on SD1.
- Become conscious of all your past lives as well as the process of rebirths.
- From now on your life will only continue to become more peaceful, more wonderful, more beautiful, more loving, and more fulfilling.
- There is no end to the wonders and bliss that you are destined to encounter.

It is my sincere hope that the material presented in this book will help the reader on the path to Spiritual attainment. It is our destiny. May peace be with you, now and forever, "even onto Nirvana, which is PEACE itself."

Bibliography

- Clinton, Hillary Rodham, *It Takes a Village: And Other Lessons Children Teach Us,* Simon & Schuster, 1230 Avenue of the Americas, New York, NY 10020. 1996
- Fox, Emmet, *Power Through Constructive Thinking,* HarperCollins Publishers, 10 East 53rd Street, New York, NY 10022, 1968.
- Hawking, Stephen, *A Brief History of Time: From the Big Bang to Black Holes,* Bantam Books, 666 Fifth Avenue, New York, New York 10103, 1990.
- Jones, Roger S., *Physics for the Rest of Us: Ten Basic Ideas of Twentieth-Century Physics That Everyone Should Know – and How They Have Shaped Our Culture and Consciousness,* Contemporary Books Inc., 180 North Michigan Avenue, Chicago, Illinois 60601, 1992.
- Kirkwood, Annie, *Mary's Massage To the World,* Blue Dolphin Publishing Inc., P.O. Box 1908, Nevada City, CA 95959, 1993.
- Kurzweil, Ray, *The Age of Spiritual Machines,* Penguin Putman Inc, New York, NY. 10014, 1999.
- McArthur, Bruce, *Your Life: Why It Is The Way It Is and What You Can Do About It,* A.R.E Press, Sixty-Eighth & Atlantic Ave. Virginia Beach, VA. 23451-0656, 1996.
- Morris, Richard, *The Nature of Reality: The Universe after Einstein,* The Noonday Press, New York, NY, 1987.

- Morris, Richard, *The Universe, the Eleventh Dimension, and Everything: What we Know and How We Know It,* Four Walls Eight Windows, 39 West 14th Street, New York, NY 10011, 1999.
- Montgomery, Ruth, *Here and Hereafter,* Ballantine Books, a division of Random House, Inc., New York, NY, 1968.
- Montgomery, Ruth, *A World Beyond,* Fawcett Book, Published by The Ballantine Publishing Group, a division of Random House, Inc., New York, NY. 1971.
- Montgomery, Ruth, *A Search for the Truth,* Fawcett Book, Published by The Ballantine Publishing Group, a division of Random House, Inc., New York, NY. 1966.
- Ramacharaka, Yogi, *Fourteen Lessons in Yogi Philosophy and Oriental Occultism,* The Yogi Publications Society, Masonic Temple, Chicago, Ill. 1903, 1904, 1931.
- Ramacharaka, Yogi, Advanced Course *in Yogi Philosophy and Oriental Occultism,* The Yogi Publications Society, Masonic Temple, Chicago, Ill. 1904, 1905, 1931.
- Ramacharaka, Yogi, *Raja Yoga or Mental Development,* The Yogi Publications Society, Masonic Temple, Chicago, Ill. 1903, 1904, 1931.
- Ramacharaka, Yogi, *Lessons in Gnani Yoga: the Yoga of Wisdom,* The Yogi Publications Society, Masonic Temple, Chicago, Ill. 1906, 1907, 1934.
- Ramacharaka, Yogi, *The Life Beyond Death,* The Yogi Publications Society, Masonic Temple, Chicago, Ill. 1909, 1937.
- Spalding, Baird T., *Life and Teaching of the Masters of the Far East, volume I through V,* DeVores & Co., 1641 Lincoln Blvd,. Santa Monica, Calif. 90404, 1964.
- Silva, Jose and Philip Miele, *The Silva Mind Control Method,* Pocket Books, a division of Simon & Schuster Inc., 1230 Avenue of the Americas, New York, NY 10020, 1977

- Van Doren, Charles. *A History of Knowledge: Past, Present, and Future,* Ballantine Books, New York, NY. 1991
- Yogananda, Paramahansa, *Man's Eternal Quest: The Collected Talks and Essays, Volume I,* Self-Realization Fellowship, 3880 San Rafael Avenue, Los Angeles, CA 90065-3298, 2000.
- Zukav, Gary, *The Seat of the Soul,* Fireside – registered trade mark of Simon & Schuster Inc., 1230 Avenue of the Americas, New York, NY, 1989.
- *The Kybalion: A study of The Hermetic Philosophy of Ancient Egypt and Greece,* By The Three Initiates, The Yogi Publications Society, Masonic Temple, Chicago, Ill. 1912, 1940.
- *The Silva Ultra Mind ESP System* – This is an audio CD cassette packet for home study, based on the *Silva Mind Control Method* and is narrated by Jonell Monaco Lytle, Michael Wickett, Jose Luis Romero and. Ed Bernd, Jr., Nightingale Conant, 7300 N. Shigh Ave., Niles, IL 60714. www.nightingale.com. Also found at Silva UltraMind, 6017 McPherson Rd. Laredo, TX 78041. www.silvaultramind.net 2000.

Printed in Great Britain
by Amazon.co.uk, Ltd.,
Marston Gate.